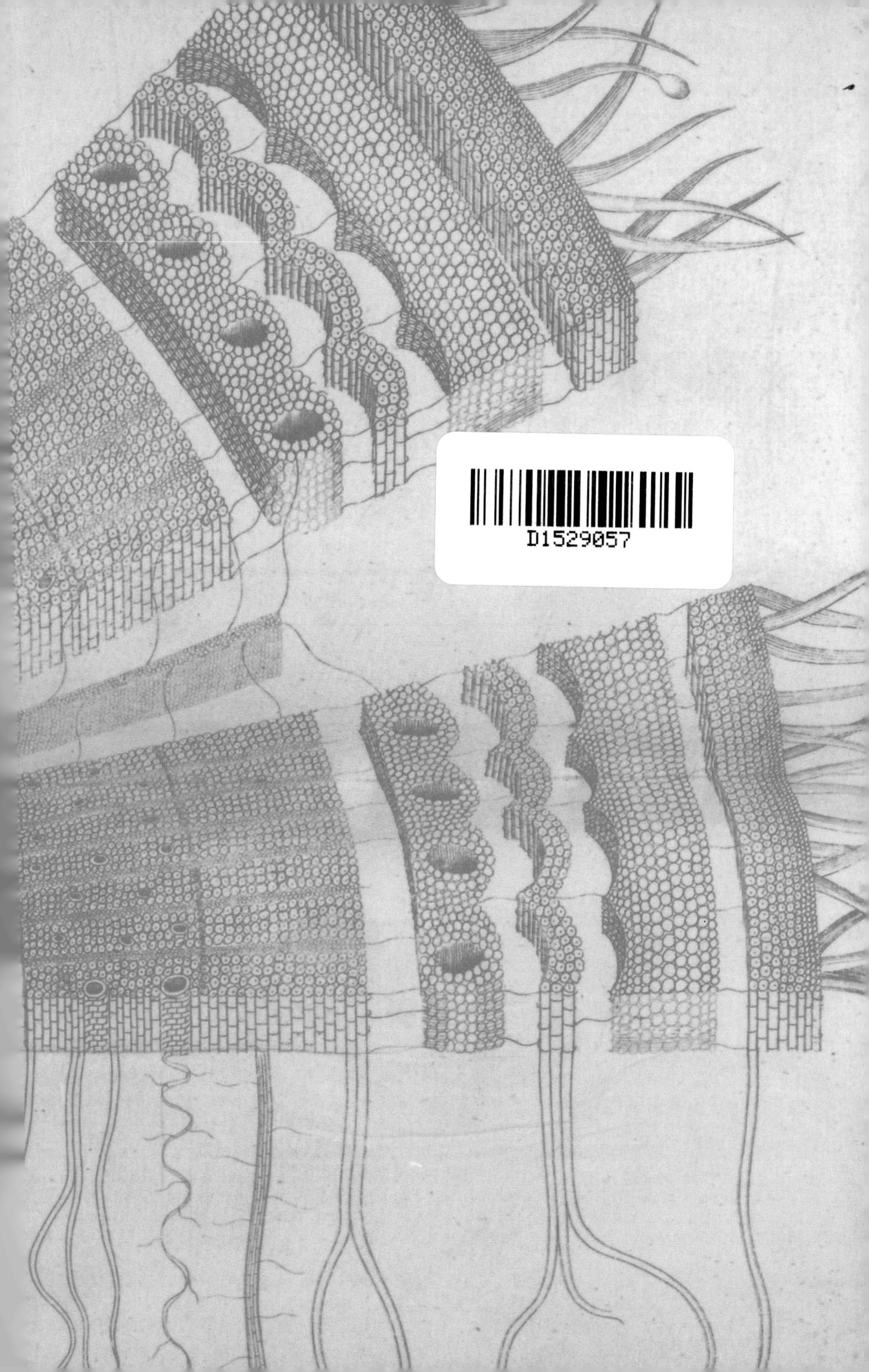

The Fabric of Life

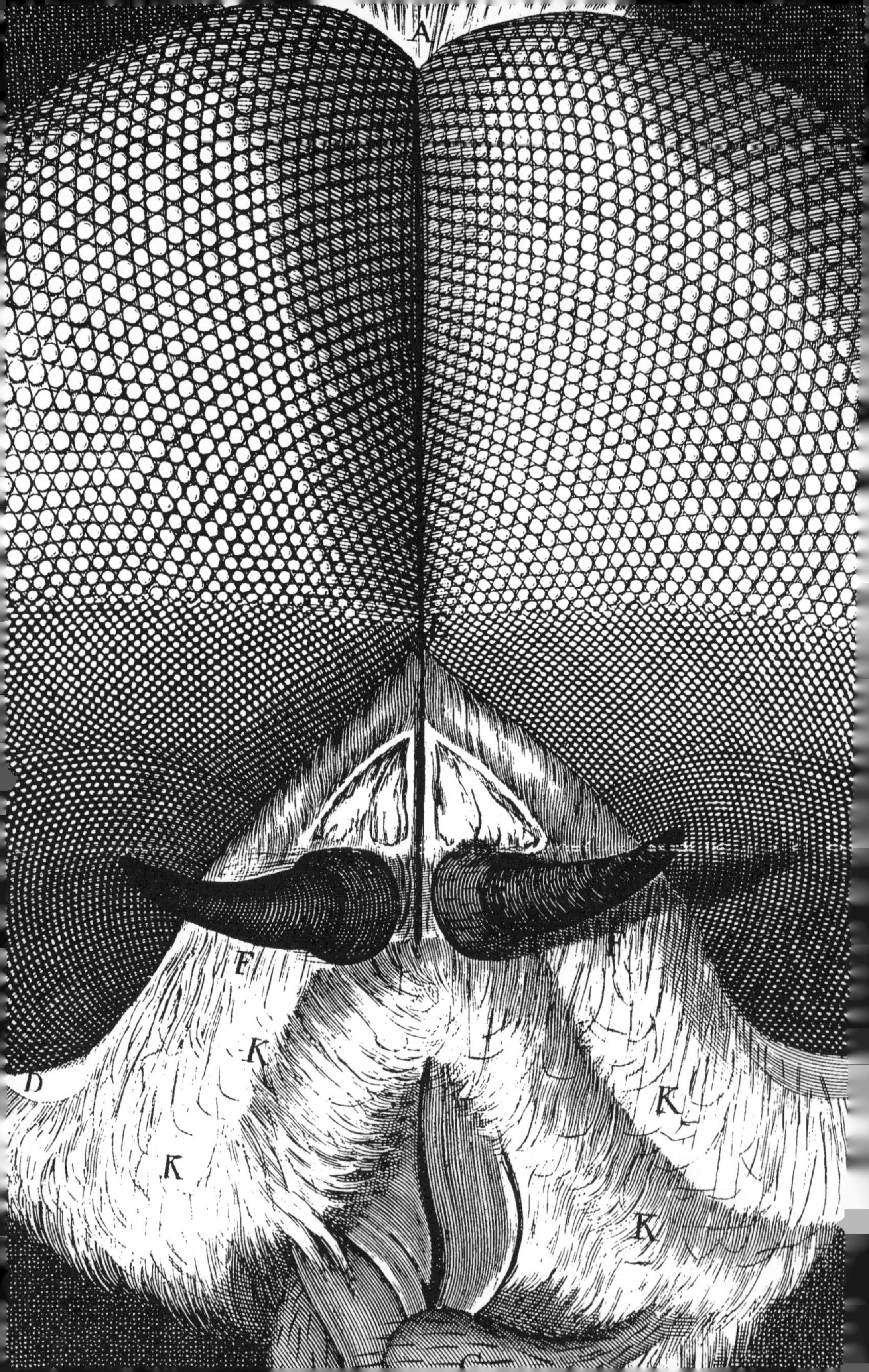
A
E
K
D
K
K
K

The Fabric of Life

Microscopy in the Seventeenth Century

Marian Fournier

The Johns Hopkins University Press
Baltimore and London

This book was brought to publication with the generous assistance of the Caecilia Stichting, the Foundation Journal Janus, and the Foundation Historiae Medicinae.

Printed in the United States of America on acid-free paper
05 04 03 02 01 00 99 98 97 96 5 4 3 2 1

The Johns Hopkins University Press
2715 North Charles Street
Baltimore, Maryland 21218-4319
The Johns Hopkins Press Ltd., London

Library of Congress Cataloging-in-Publication Data
will be found at the end of this book.

A catalog record for this book is available
from the British Library.

ISBN 0-8018-5138-6

Contents

Acknowledgments vii

Introduction 1

1. A New Instrument Appraised 9

2. The Leading Microscopists 49

3. The Substance of Living Matter 92

4. The "Animal Oeconomy" 104

5. The Fabric of Living Beings 135

6. Five Heroes of Microscopic Science 178

7. Measuring the Impact of the Microscope 185

Conclusion 197

Appendixes

A: Books on Microscopy Published between 1625 and 1750 201

B: Articles on Microscopy in the *Journal des sçavans* between 1665 and 1750 203

C: Articles on Microscopy in the *Philosophical Transactions* between 1665 and 1750 203

D: Articles on Microscopy in the *Miscellanea curiosa* between 1670 and 1750 210

Notes 215

Bibliography 237

Index 261

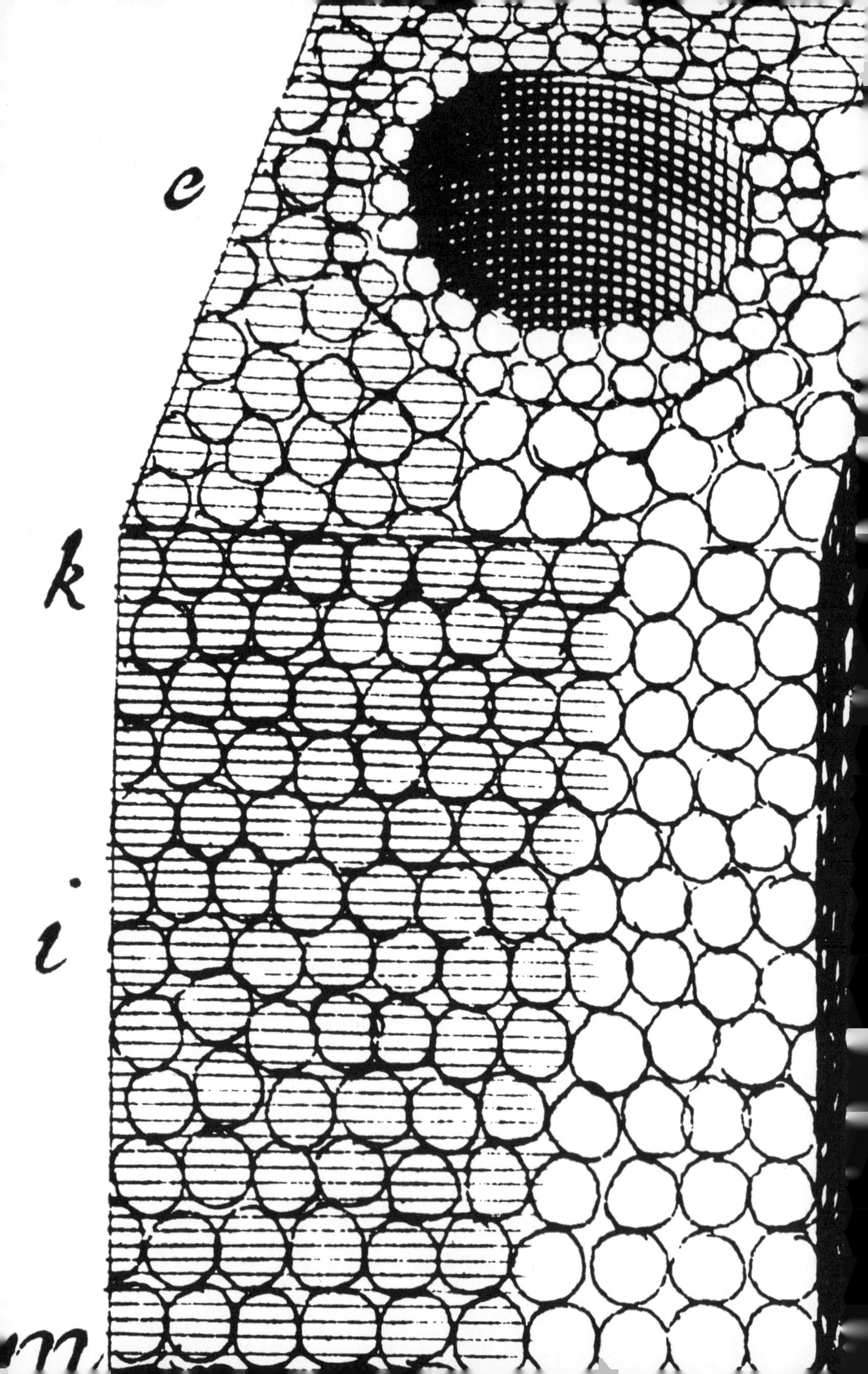
e
k
i

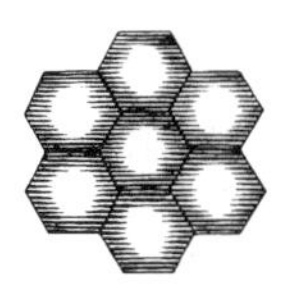

Acknowledgments

THE PLAN FOR this book originated in the preparations for the Antoni van Leeuwenhoek exhibition that was held in the Museum Boerhaave in 1982 and in the Science Museum (London) in 1983. This fact symbolizes the ambience of the Museum Boerhaave: an intimate connection between the care for historical scientific apparatus and the scholarly investigation of the history of science. It is undoubtedly due to the director of the Museum Boerhaave, Dr. G. A. C. Veeneman, that this link has strengthened and flourished during the past decade. I feel grateful to him both for the creation of a constructive atmosphere to work in and for his personal support during the years in which I prepared this study.

The writing of this book has greatly benefited from the help of my colleagues at the Museum Boerhaave. Their support ranged from discussions about its substance, through typing the earliest drafts and collecting photographs, to making the computer perform various productive feats. I am especially indebted to the librarian of the museum, H. J. F. M. Leechburg Auwers, for his invaluable assistance in preparing the appendixes and bibliography.

The present book is a somewhat altered version of my doctoral thesis, which was defended at Twente University (Enschede) in 1991. During the years I prepared the thesis I greatly appreciated the constant support of my supervisors, Prof. Dr. H. F. Cohen and Prof. Dr. P. Smit, who accepted that sometimes museum affairs had to take precedence over the completion of my dissertation. I also appreciate the valuable comments of Dr. B. Bracegirdle, Dr. W. Th. M. Thijssen and Dr. J. A. M. Kuylen, who read earlier drafts. Prior to rearranging my thesis into the present book I discussed its outlines with Dr. R. P. W. Visser, whose suggestions are gratefully acknowledged.

The sections on Jan Swammerdam have appeared in the second volume of *Tractrix* and were included in my dissertation with the permission of the editors of that journal.

I am deeply indebted to the Museum Boerhaave for supplying and giving permission to publish all the illustrations included in this book.

The Fabric of Life

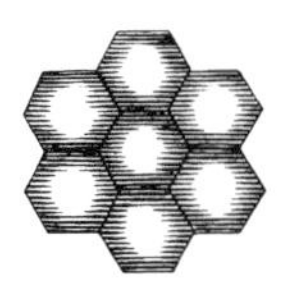

Introduction

THE MICROSCOPE is today a highly valued instrument for scientific investigation. Without its help large areas of research in medicine, biology, and other branches of science would be virtually impossible. It may be surprising to learn, however, that the appreciation of this splendid instrument did not arise immediately after it was invented.

The development of the microscope was a direct result of the invention of the telescope, which happened around 1610 in Middelburg, the capital of the Dutch province of Zeeland. Names that are in one way or another connected with its development are those of Zacharias Jansen, Hans Lipperhey, and Jacob Metius. In contrast to the telescope, which was within a year or two of its invention used by Galileo Galilei to examine the moon and the night sky, resulting in one of his most celebrated discourses, the microscope did not attract much attention nor were its possibilities fully recognized until some fifty years later. Only in the early 1660s and in a remarkably short span of time did the microscope secure a great reputation, and many scholars came to expect that this instrument was one of the finest means to discover the deepest secrets of nature.

In 1664 Robert Hooke voiced his hopes for the future of the microscope in the preface to his famous book *Micrographia* as follows:

> It seems not improbable, but that by these helps [i.e., the microscope] the subtility of the composition of the Bodies, the structure of their parts, the various texture of their matter, the instruments and manner of their inward motions, and all the other possible appearances of

> things, may come to be more fully discovered. . . . From whence there may arise many admirable advantages, towards the increase of the *Operative,* and the *Mechanick* Knowledge . . . , because we may perhaps be inabled to discern the secret workings of Nature.[1]

Clearly Hooke looked forward to a productive period of microscopic research, a period in which that instrument would help him, and many others as well, to unravel some of the greatest mysteries of nature. Barely thirty years later, however, after reading a paper to the fellows of the Royal Society in 1692, he deplored the current neglect of the telescope and observed discontentedly:

> Much the same has been the Fate of Microscopes, as to their Invention, Improvements, Use, Neglect and Slighting, which are now reduced almost to a single Votary, which is Mr. *Leeuwenhoek;* besides whom, I hear of none that make any other Use of that Instrument, but for Diversion and Pastime. . . . If we enquire into the Reason of this Change of Humour, in Men of Learning, in so short a Time, we shall find that most of those, who formerly promoted these Enquiries, are gone off the Stage; and with the present Generation of Men the Opinion prevails, that the Subjects to be enquired into are exhausted, and no more is to be done.[2]

Presuming that in the first quotation Hooke voiced a generally shared expectation, and that in the second he gave an accurate analysis of the prevailing attitude toward microscopy, it is apparent that within only three decades the microscope proved to be something of a disappointment rather than an invaluable tool. In the early 1660s it was expected that some of nature's hidden secrets concerning the structure and function of living beings would be discovered. These expectations were fulfilled to some extent, as there were men who "promoted these Enquiries." However, the interest of investigators faded within a few decades, so that toward the end of the seventeenth century only one man—Antoni van Leeuwenhoek—continued to use the microscope for the investigation of nature.

It is universally accepted in the history of the life sciences that microscopy was in the front line of research during the second half of the seventeenth century. During this period an extensive body of data concerning the microworld was discovered, among these the discovery of the capillaries of the vascular system, the plant cell, the spermatozoa, and the unicellular organisms stand out. These discoveries were then—and now—widely appreciated as being among the most momentous of the times, both within the field of the life sciences and the domain of the physical sciences. Most of the outstanding findings were contributed by only five men: Robert Hooke, Marcello Malpighi, Jan Swammerdam, Nehemiah Grew, and Antoni van Leeuwenhoek. A fairly large number of scholars of lesser repute were encouraged by the amazing discoveries reported by these five to follow their lead, and they too added considerably to the body of common knowledge.

Although the geographical distance between the five most prominent men was substantial (Hooke and Grew worked in London, Malpighi in Bologna, Leeuwenhoek in Delft, and Swammerdam in Amsterdam), they were conversant with each other's work, sometimes only through reports from a correspondent, but more often because they had read each other's books. It is a remarkable and intriguing fact that the Royal Society played an important role in the careers of the leading microscopists: Hooke's excursion into the microworld was more or less commanded by his fellows of the Royal Society. Grew, Malpighi, and Leeuwenhoek were urged by the members of the Royal Society to continue and expand their investigations, and that body also discussed in the course of several of its weekly gatherings the results of the investigations of these men and advanced the publication of their scientific works.

A rapid survey of the relevant literature suggests that the decades of flourishing microscopic research coincided with a marked development of the microscope.[3] In the decades prior to 1660 few or no changes were made in its construction. In fact, during that period only a single improvement to the microscope is reported, the addition of the field lens, which proved in practice to be a rather dubious improvement. After 1660, on the other hand, a variety of microscopes of different designs and novel constructions were produced by various instrument makers throughout Europe, and a succession of suggestions for further improvements, including alternative optical arrangements, are found in contemporary publications. Apparently the regular use of the microscope greatly encouraged its further development.

In most historical surveys of the life sciences the fact that by the beginning of the flourishing of microscopy ca. 1660 the microscope had already been in existence for nearly half a century, without concomitant scientific application, is usually barely recognized. Yet the sudden change of scholars' attitudes toward this instrument from one of apparent indifference to one of appreciation after it had been available for four decades suggests that some specific stimulus brought about this change, a stimulus that promoted the prompt employment of the microscope in scientific exploration.

The rise to prominence of the microscope began more or less at the same time that the appreciation of the mechanical philosophy gained momentum. Several historians of science have argued that the introduction of the mechanical philosophy into the life sciences was the principal catalyst of the rise of microscopy.[4] Within the framework of this novel system of thinking about the natural phenomena, the operation of organic matter was assumed to transpire through a variety of minute, but distinctive—and therefore visible—structures. The ensuing investigations disclosed in a very short span of time a wealth of new particulars concerning the anatomy of animals and plants, which established that the complexity of structure of the various organisms was much greater than anyone had ever dreamed and, even more spectacularly, a whole new world of heretofore undiscovered organisms materialized from beneath the surface of the waters.

Although a fairly large part of the microscopic studies of this period were manifestly aimed at the explication of physiological processes in mechanistic terms, an equally large, if not larger, number of investigations focused on purely anatomical topics and the details of the natural history of animals and plants. All in all, scholars, whether they themselves applied the microscope or not, were faced with the challenge of explaining a mass of unexpected findings and fitting them into the general body of learning.

It is also generally maintained in the history of science that microscopic research was at a very low level during the greater part of the eighteenth century. However, a careful examination of contemporary publications shows that the microscope was in fact widely used in natural history by such men as René-Antoine de Réaumur, Pier Antonio Micheli, Charles Bonnet, Luigi F. Marsigli, Pierre Lyonet, Abraham Trembley, and John Ellis. Among them they studied sundry insects, the spores of fungi, the "flowers" of coral, the fresh water polyp, and corallines. At the same time the microscope became a very fashionable gadget among the well-to-do. Large numbers of microscopes found their way to the homes of "lovers of science" and were regularly used for the inspection of multifarious organic objects. The breathtaking complexity and variety of organic nature formed a welcome subject for the studies of the average amateur of science, proving to his admiring eyes beyond question the omnipotence of God.

Simultaneously, the design and construction of the microscope underwent many changes, resulting by the end of the eighteenth century in the establishment of the pillar microscope as the principal design. Moreover, a variety of books were published that introduced the serious amateur to microscopic techniques and suitable objects for study.[5] The outstanding microscopic discoveries of the preceding century were certainly accepted by their eighteenth-century successors and were assimilated into the common body of knowledge. Microscopic fibers played an important role in theories about the construction and operation of living matter throughout the eighteenth century, and the vast number of different kinds of infusoria was carefully itemized in the course of that age.

The eighteenth century was therefore not a period in which the microscope was no longer used: on the contrary, many explorers of nature used a microscope, either for serious scientific observations or to savor edifying images. The observations of these students of nature, however, no longer had a significant bearing on the development of the pivotal scientific ideas of the eighteenth century. What then had been the reason to banish the microscope from the front line of scientific discovery to the rearguard? Admittedly, scholars were disappointed and frustrated because, looking through their instruments, they so often only caught tantalizing glimpses of details that they could not resolve with the instruments at their disposal. Malpighi, for instance, deplored at times the poor quality of contemporary microscopes. Even so, the researches of the above-mentioned eighteenth-century scholars demonstrate that new areas of research were opened with microscopes that were only slightly—if at all—superior to those used by their predecessors.

This book focuses on the question why microscopic research apparently prospered for only some four decades in the seventeenth century. From the outset it is assumed that the rise and subsequent decline of microscopy within so short a span of time were somehow linked. Such a thesis implies that the original goals for applying the microscope to study nature's smaller structures were in some way defeated by those very investigations.

The current perspective on the flourishing of seventeenth-century microscopy is insufficient in several respects. The beginning of this new technique is attributed to the simultaneous rise of the mechanical philosophy simply on the evidence of the contents of a part—and those the best known of a larger number—of the microscopic investigations without accounting for the numerous observations on all sorts of vegetable and animal subjects, which bear no relation to the issues raised in the context of the mechanical philosophy. Also, it is too readily accepted that the microscope was barely employed in science in the eighteenth century, thereby neglecting the sound microscopic investigations in natural history. It is the aim of this study to assess the influence of the mechanical philosophy on the rise to prominence of the microscope more precisely, to establish whether or not the microscopic observations in natural history were inspired by the same tradition and to discover why the microscope was set aside again so quickly. The answers to these questions are of considerable importance since they highlight the relevance of the mechanical philosophy for the fundamental issues of the life sciences in the seventeenth and early eighteenth century.

The microscope and microscopy in the seventeenth century are by no means neglected subjects in the history of science. Indeed, the improvement of the microscope, its makers and users, and the discoveries made with it have been studied by historians of science from various viewpoints. By far the widest attention has been devoted to the development of the microscope, which has been studied in great detail, particularly the development of its design and mechanical construction.

The accomplishments of each of the leading microscopists of this period have been duly recorded and valued. A large number of the relevant studies are biographical and concentrate on enumerating the contributions of individual microscopists. Other studies trace the unique development of the ideas generated by the microscopic investigations of one particular scientist in relation to the whole of his scientific output. From a cursory examination of the various titles of these studies it appears that historians of science tend to center on the discoveries that have proved to be of lasting value, such as Leeuwenhoek's discovery of the infusoria and Malpighi's discovery of the capillaries of the vascular system. Furthermore, the impact of microscopic investigations on contemporary life sciences is assessed in a variety of studies concerned with the development of particular biological theories or disciplines. The bulk of historical studies on microscopy have therefore been devoted to either a particular microscopist or to a given subject in the life sciences. In other words, these studies were concerned with a limited part of microscopy, infrequently taking current developments into account.

An inquiry into the goals and efforts of the principal investigators of the period—Hooke, Malpighi, Swammerdam, Grew, and Leeuwenhoek—forms a substantial part of the present appraisal of the changing fortunes of the microscope. The publications of these men were major events, not only in the eyes of posterity, but also in the estimation of their contemporaries. Since the endeavors of the leading microscopists had a bearing on some of the most seriously debated issues in contemporary science, their peers naturally joined them in the discussion of these issues and investigated some of the subjects they had initiated. A survey of the investigations carried out by second-rank microscopists and an analysis of the impact of microscopy on contemporary life science therefore supply the necessary context for the achievements of the principal microscopists. Joined together, the inquiries into the scientific commitments of the prominent and less prominent microscopists, the substance of their explorations with the help of the microscope, and the impact of their discoveries on contemporary science may be expected to provide a comprehensive view of the incentives underlying this peak period of microscopy.

The present book therefore differs considerably from the existing body of historical writing on microscopy: the investigations of the principal microscopists and of a good number of microscopists of lesser fame are united into one inclusive story—a story that is primarily based on the assessment of the microscopic findings and the growth of scientific thought, wherefore the perspective is resolutely internalistic. Several issues that clearly have influenced the development of microscopy (the role of scientific societies is a case in point) are not, or are only very superficially, discussed. In addition, the impact of microscopic discoveries on late-seventeenth-century society (which must have been profound, at least in England, considering the number of plays ridiculing microscopy put on stage in London),[6] and the opportunities this instrument proffered young—and not so young—ladies to participate, however far removed, in the adventure of scientific discovery are subjects that will not be addressed.

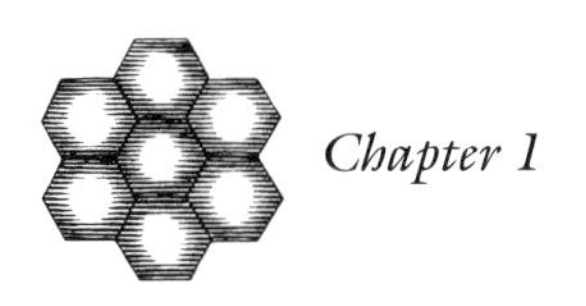

Chapter 1

A New Instrument Appraised

MICROSCOPIC INVESTIGATION flourished in the second half of the period that is commonly known as the Scientific Revolution. Although this designation is often equated with the profound changes that were brought about in the explanatory framework of science, it is also used, and as such it will be used throughout this book, to indicate merely a period of great changes and rapid development in the sciences. One of the most notable changes was the increasing use of instruments to measure, manipulate, and experiment with natural phenomena. Whereas by the beginning of the sixteenth century few instruments besides drawing instruments and simple devices for measuring lengths, angles, and weights were available, a century later scientists could choose from a variety of instruments, including the telescope, microscope, air pump, thermometer, and clocks, to investigate nature.

The Development of the Microscope

Reputedly, Galileo Galilei was the first to use a compound optical device to magnify a small object in 1609 or 1610.[1] This device was not what is now known as a microscope, but rather an early version of the telescope, the Galilean or Dutch telescope. This consisted of a combination of a convex and a concave lens, fitted into an extensible tube, which could be used to magnify either far-off things or small objects.

The invention of the compound optical tube, or "perspicillum" as it was then often called, cannot with certainty be attributed to any one person or dated exactly. The consensus of opinion on this historical event is that, ca. 1600, spectacle makers sometimes used combinations of convex and concave lenses to remedy certain ophthalmological defects, a practice that prepared the way for the invention of the Dutch telescope in which the same combination is used, although with stronger lenses.[2] One day someone grasped the significance of the effect he observed when chancing to line up just such a combination of lenses.

The oldest extant document relating to the invention of the telescope is a letter bearing the date 25 September 1608 from the Committee of Councillors of the States of Zeeland in Middelburg, a town in the Netherlands.[3] The letter was intended to support an application for a patent by Hans Lipperhey, a spectacle maker in that town, with the States-General of Holland, allowing him to manufacture this device. Lipperhey's request was not granted, however, because the authorities learned (partly through a similar request by Jacob Metius, reaching them only a few days later) that others also knew of the invention or were involved in it.

Although the inventor of the telescope will probably be forever unknown, there is no doubt that Galileo introduced this instrument into scientific practice. After a rumor about the existence of this new device had reached him, Galileo worked out how it should be constructed, and promptly acted on that knowledge. As early as 1610 his *Sidereus nuncius* was published in which he talked about the observations of the moon and other heavenly bodies made with his "perspicillum."

Galileo's book had an enormous impact on the scientific community. Johannes Kepler, for one, wasted no time in exploring the optical properties of combinations of lenses and published these studies in his *Dioptrice* of 1611. From his studies it appeared that the telescopic effect could also be attained by a combination of two convex lenses. Such a system produces a larger and brighter image, albeit upside down. The combination of convex lenses in an optical system became known as the Keplerian arrangement and had by 1640 almost replaced the earlier Dutch arrangement.

Cornelius Drebbel's name is firmly linked with the advent of the first compound microscope—that is to say, one with a Keplerian optical system. Although the first mention of an actual microscope is linked with his name as the owner, this does not inevitably mean that he also invented it. It is certain, however, that in the early 1620s several microscopes made by Drebbel appeared all over Europe. For instance, in 1622 Jacob Kufler, a brother of Drebbel's son-in-law, presented a microscope to Marie de Medici, Queen of France. Attending her court was Nicolas-Claude Fabri de Peiresc, who acquired several copies for his own use. Through Kufler and Peiresc several copies were presented to representatives of Roman notables in 1623.[4] Galileo came across these copies in 1624 when his help was requested by the owners who could not make them work even with

Peiresc's detailed instructions. Galileo then proceeded to construct similar microscopes and presented some examples to Prince Federico Cesi, president and founder of the Accademia dei Lincei. Within this body of learned men the microscope first came into its own: the earliest microscopic researches were published by some of its members, namely, a study of the bee by Federico Cesi and Francesco Stelluti. Moreover, the microscope was given its name by one of them: Johannes Faber mentioned in a letter to Prince Cesi, dated 13 April 1625, Galileo's new optical device for investigating small things, calling it a microscope.[5]

From extant sources it appears that these earliest microscopes consisted of a three-piece extensible brass tube.[6] At the objective end, the tube was fitted with a small plano-convex lens, the convex side of which was turned toward the object. Toward the other, ocular, end of the tube a much larger, double convex lens was situated some five centimeters ("deux doigts") from its mouth. The tube was encased in a brass ring, fitted with three legs that were fastened to a small disc. In the middle of the disc a stage was mounted. Focusing could be achieved by sliding the tube up or down in the tripod stand, while variable magnification could be accomplished by drawing out the tube to different lengths.

Quite a few elated assessments of the "fly the size of an elephant" kind may be found in early reports on the performance of the microscope. Such rapturous praise, however, became mingled during the second half of the seventeenth century with keenly felt chagrin caused by the limitations of the optical part of the microscope. The inadequate images produced by microscopes were partly due to the poor quality of the glass that was used, which often suffered from various flaws such as air bubbles and small impurities. Moreover, the grinding of lenses was a difficult task requiring patience and craftsmanship. Although the invention of the telescope originated from the craft of spectacle making, which of course included the grinding of lenses, this craft did not rise to the occasion. The manufacture and further development of optical instruments was left to the relatively novel instrument-making trade.[7] This new metier had to master the necessary skills before it could respond to the demands imposed on it by the invention of optical instruments. Consequently it was not until the end of the century that commercially produced microscopes became available in any quantity.

Even more important than the quality of the lens glass were the limitations imposed by the laws of refraction. Spherical lenses distort the image because the refraction at the edge of the lens is greater than in the center, causing a blurring of the image known as spherical aberration. Another, equally serious, problem was due to the varying degrees of refraction of light of different colors. Chromatic aberration could become particularly serious in compound microscopes, where the light is refracted several times, the result being colored fringes around the image. Optical theory of that time held that spherical aberration could be avoided by employing aspherical lenses, but these proved impossible to grind. Chromatic aberration, on the other hand, could not be avoided by changing the form of the lens. This was demonstrated by Isaac Newton who, in an article in the *Philosophical Transactions* of 1672, held that chromatic aberration could not be corrected at all. In those days the pragmatic solution to the problem of spherical aberration was to fit a diaphragm over the objective lens so that only the middle part of the lens was used. This procedure, however, meant the loss of a considerable amount of light, so that the image became indistinct. The instrument maker therefore had to find a middle way between reducing spherical aberration by stopping down the objective, and losing distinctness altogether.

With a view to improving its performance and reducing its imperfections, the elementary optical arrangement of the compound microscope was subjected to a number of additions and alterations, beginning with the introduction of the field lens about 1650. The field lens, situated between the objective and ocular, widened the field of vision and also moved the point where the eye of the observer had to be in order to view the image to a more convenient position much nearer to the ocular. This third lens was introduced simultaneously in telescopes and microscopes by the instrument maker Johannes Wiesel of Augsburg.[8] However, by adding a third lens to the system the overall results of the various lens defects became even more trying. Hooke pointed out the various advantages of a system with and without a field lens. He wrote, "This [the field lens] I made use of only when I had occasion to see much of an Object at once; . . . But when ever I had occasion to examine the small parts of a Body more accurately, I took out the middle Glass, and only made use of one Eye Glass with the Object Glass, for always the fewer the Refractions are, the more bright and clear the Object appears."[9]

The introduction of an alternative ocular lens by Eustachio Divini in 1667 constituted another optical development. Divini produced eyeglasses consisting of a doublet of two plano-convex lenses fitted together in such a way that the convex side of each lens was turned to the other, just touching in the middle. He claimed that this arrangement of lenses displayed the objects without distortions and yet magnified extremely well.[10]

Even though the addition of lenses to the optical system tended to have deleterious effects on the quality of the image, many scientists and artisans experimented with even more complicated systems, particularly by introducing doublets. Christopher Cock, for instance, made a microscope for the Royal Society in 1669 consisting of no less than five lenses, four of which made up the ocular.[11]

Some twenty years later Johann Franz Griendel von Ach designed an optical system containing three doublets, each consisting of two plano-convex lenses, with the convex sides turned toward each other. These doublets functioned as ocular, field lens, and objective, respectively. This arrangement, he claimed, doubled the field of view and permitted a greater distance between the object and objective, thereby facilitating illumination and approach of the object.[12]

Despite these efforts, the one substantial improvement in the performance of the microscope was made by reducing the number of lenses, now relying on a single lens only—the simple or single-lens microscope. Magnifying glasses, flea glasses, and similar optical instruments, perhaps best characterized as simple microscopes of minor strength, had of course been in use since the beginning of the seventeenth century. However, the simple microscope of 1700 was an altogether different affair: it was fitted with several gadgets that ensured easy handling and, even more important, it possessed far greater optical powers.

During the early 1660s several people, realizing that one tiny spherical lens produced greater magnification and also suffered far less from chromatic aberration, experimented with the grinding of such lenses. Johannes Hudde, for instance, a promising mathematician who in later years preferred an administrative career, corresponded with Christiaan Huygens about the advantages of the simple microscope.[13] Hooke related in his *Micrographia* how he had experimented with these lenses and had found the handling of them very difficult and also "offensive to his eyes." Nevertheless, he conceded that "'tis possible with a single Microscope to make discoveries much better than with a double one, because the colors which do much disturb the clear vision in double Microscopes is clearly avoided in the single."[14] The advantages of the simple microscope were obvious, and this instrument was therefore frequently used in superior microscopic investigations, such as those executed by Antoni van Leeuwenhoek, who mastered to perfection the art of making microscopes with miniature lenses.

Although quite a few data concerning various aspects of the microscope, among them statements as to their magnification, may be accumulated from a variety of published sources and manuscript material, not a single artifact dating unquestionably from before 1670 has survived the vicissitudes of time. Therefore, any judgment concerning the performance of the earliest microscopes must be uncertain.

Before 1700 few statements concerning the magnification of a microscope were expressed in figures. In 1654 Huygens estimated the magnification of a compound microscope in his possession to be fifty-six times.[15] Divini asserted in 1667 that a microscope made by him, and fitted with a doublet ocular, could magnify 41, 90, 111, or 143 times, depending on the distance between objective and ocular.[16] Griendel von Ach claimed in 1687 that the microscope he had developed magnified about a hundred times.[17]

Another contemporary source of information on the strength of microscopes is the illustrations added to the publications. Stelluti's drawing of the bee (figure 1), Odierna's illustration of the fly's eye (figure 2), and a crude drawing of a moth with an enlargement of one of its antennae, from one of Borel's hundred microscopic observations (figure 3), give some indication of the powers of microscopes prior to 1660. Obviously these microscopes did not magnify very much: in Stelluti's and Borel's case less than ten times, and in Odierna's no more than twenty or thirty times.

Several historians of science have actually measured the optical performances of artifacts dating from the late seventeenth and eighteenth centuries.[18] As they employed different methods to arrive at their data and have also determined different parameters, their results cannot be directly compared. However, a consensus of opinion on the question is easily extracted from their conclusions. Between 1670 and 1750 the performance of both the simple and compound microscopes did not improve to any great extent. The microscopes produced by one instrument maker exhibited great variation in the quality of the image. The images rendered appear inferior to modern eyes, in that they suffered severely from chromatic and spherical aberrations, giving an indistinct and limited view. The simple microscope, as expected, scored better in these respects. The compound microscope of this period uniformly had a small aperture: the best magnified up to one hundred times, but the majority of them only up to fifty, with a resolution of at most ca. 3 microns. Most instrument makers generally supplied the simple microscope with a set of interchangeable lenses after about 1700. A Wilson-type microscope would, in a set of five lenses, on average cover a range of 25–150 times, with resolutions up to 2.5 microns. Therefore, throughout the period under consideration, the simple microscope was a superior instrument compared to the compound microscope.

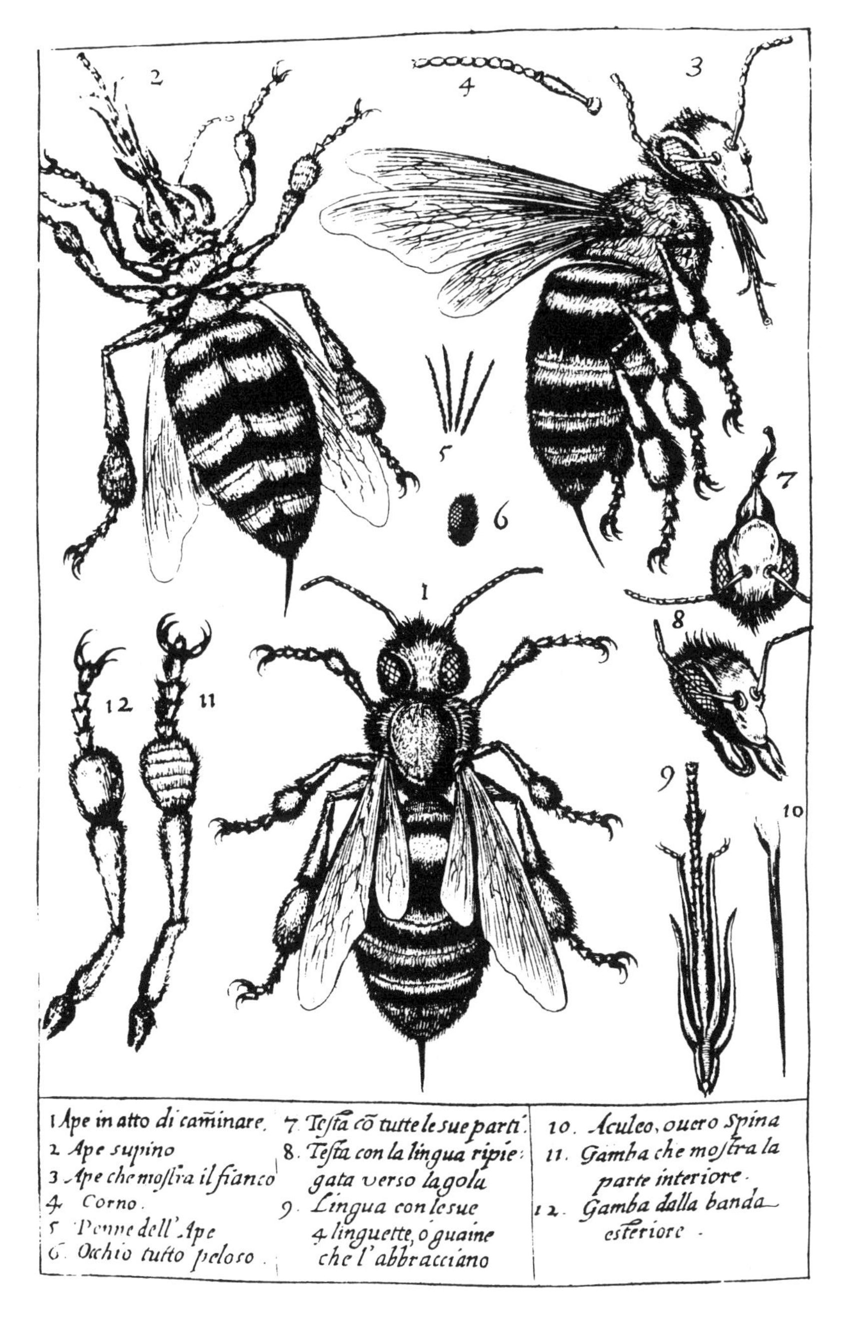

Fig. 1. Stelluti's representation of the exterior of the bee and some of its parts, from his *Persio tradotto* (1630).

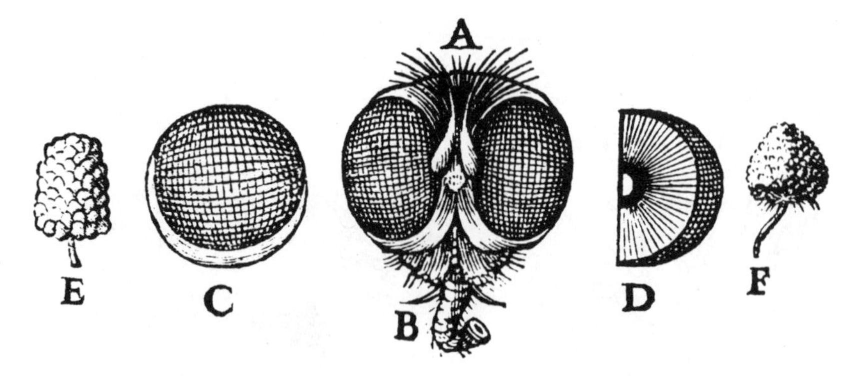

Fig. 2. Odierna's illustration of the exterior (A, C) and a cross section (D) of the compound eye of the fly, from the title page of his *L'Occhio della mosca* (1644).

Fig. 3. Borel's drawing of a moth with an enlargement of one of the antennae, from his *Observationum microscopicarum centuria* (1656).

The Leeuwenhoek microscopes have been held up in the history of science again and again as models of perfection and high quality. Recent scholarship has revealed that Leeuwenhoek's fifteen extant microscopes and lenses, comprising thirteen lenses, have a magnification varying between 3 and 266 times.[19] The majority, twelve to be exact, range from 30 to 167 times. The resolving power of these twelve lies between 4 and 2 microns with a lower and upper limit of 8 and 1.35 microns, respectively. From a report on a number of Leeuwenhoek's microscopes, which have since vanished, it appears that these microscopes magnified between 50 and 200 times.[20] From these data we can determine even more precisely that the majority of Leeuwenhoek microscopes magnified between 75 and 150 times.

Comparing these facts about the Leeuwenhoek microscopes with the data on commercially produced microscopes, it is obvious that Leeuwenhoek produced, among the 350 or so microscopes he made in the course of his career,[21] a small number of really outstanding microscopes, which had much greater resolution than any others dating from that time. However, it is equally clear that as soon as the instrument-making trade had caught up with the required technology for grinding lenses, simple mi-

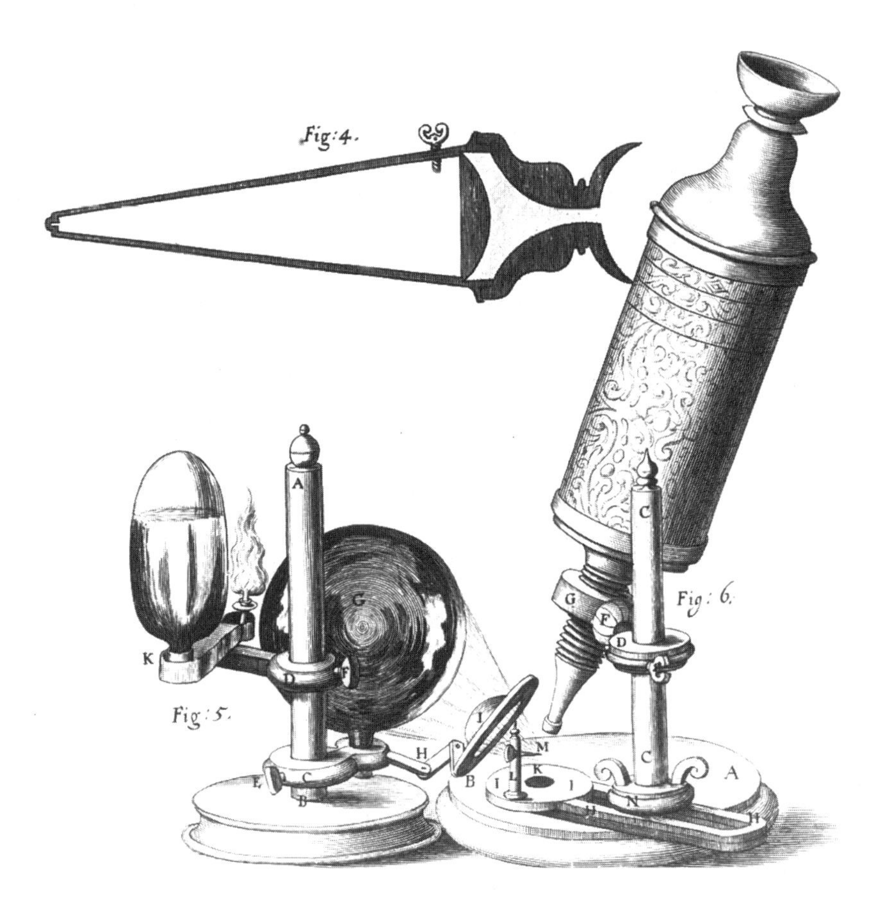

Fig. 4. Hooke's microscope as illustrated in table 1 of his *Micrographia* (1665). The illustration includes the exterior of the microscope as well as a cross section, showing the position of the lenses (fig. 4) and a lamp Hooke used for illuminating his preparations.

croscopes equal in quality to Leeuwenhoek's average microscopes could be acquired commercially.

The earliest design of the microscope, an extensible tube fitted with lenses and encased in a tripod stand, formed the basis for further developments. The focusing mechanism of the tripod stand, sliding or screwing the tube up or down along the collar of the stand, proved to be rather crude and inconvenient. The single-pillar stand (figure 4), introduced by Hooke in his *Micrographia* in 1665, allowed for easier and more accurate focusing in two stages. Overall focusing was brought about by clamping the collar, which supported the tube, at approximately the right position along the pillar, while fine adjustments could be made by screwing the tube up or down in the collar. The single-pillar stand was adopted by the London instrument maker John Marshall, who modified the focusing mechanism in agreement with a proposal made by Johannes Hevelius. In 1673 Hevelius had designed a focusing screw consisting of two parts,[22] the upper part being a sleeve that could be clamped to the pillar. This sleeve was connected by means of a screw-threaded rod to a second sleeve supporting the body of the microscope. Throughout the eighteenth century both mechanisms were employed. The Italian instrument maker Giuseppe Campani and the Englishman Edmund Culpeper used, respectively, the screwing and sliding mechanisms. Marshall's design on the other hand was adopted and improved on by his compatriot John Cuff in 1745.

Hooke's design featured some other novelties. He had inserted a ball-and-socket joint between the ring supporting the tube and the clamp, so that the tube could be tilted at any angle. This arrangement was, however, susceptible to wear and inadvertent movements; Marshall therefore fitted the ball-and-socket joint at the bottom of the pillar, which achieved the same overall effect without those disadvantages.

The mechanical stage was another of the facilities introduced by Hooke. Whereas earlier microscopes were equipped with a simple stage on which the object was laid out, Hooke's mechanical stage allowed the object to be studied from various angles without removing it from (microscopic) sight. The mechanical stage, however, although incorporated into the Marshall microscopes, was not widely employed until the second half of the eighteenth century.

The construction of the early compound microscope was suited to observation of opaque objects with incident light only. Once the attention of scientists and craftsmen became focused on the microscope, its range of action was extended by adding suitable provisions for observation with transmitted light. Initially the microscope was simply lifted and directed toward a bright light, the sun or the flame of a candle, obviously a rather inconvenient method. In 1679 Christiaan Huygens suggested the use of a substage mirror to his brother Constantijn,[23] but this practical device was not adopted until thirty years later. In 1712 Christiaan Gottlieb Hertel described in his *Novem inventuum microscopii . . . descriptum* an elaborately worked microscope fitted with a substage mirror. A few years later Culpeper seized on this improvement and always fitted his microscopes with a mirror of this kind, setting an example that was to be universally followed by other instrument makers. The design of Culpeper's microscopes, which became famous and was widely imitated, was essentially the same as the original tripod stand, to which he added a second tripod in such a way that the original base plate became the stage, mounted over the mirror (see figure 5).

In the 1680s the construction of the microscope attracted much attention in Italy, resulting in several novel ideas. Giuseppe Campani, for instance, introduced a device for wedging an object slide in place.[24] This device consisted essentially of two brass plates, pierced with wide central holes and firmly clasped together by two spring clips, attached to the lower plate. Campani's idea was taken up by Marco Celio, a member of the Accademia Fisicomatematica Romana, who developed it into the spring stage. The spring stage bears great resemblance to Campani's stage, but the spring clips are replaced by a spring pressing the two plates together. This construction was finally made known to scientific circles through Philippo Buonanni's *Micrographia curiosa,* published in 1691. The spring stage became very popular and was widely employed throughout the eighteenth century. As a result of these developments the average eighteenth-century microscope was delivered to the customer in a case, supplemented with a number of accessories such as a set of interchangeable objectives, spring stage, loupe, fish plate, and small preparation tools.

Once Leeuwenhoek's microscopic researches began to be published, interest in the simple microscope was stimulated, and consequently instrument makers and scientists directed their attention toward the improvement of this instrument as well. Leeuwenhoek's microscopes comprised magnifying glasses, aquatic microscopes, and the simple microscope of his own design (see figure 6). The aquatic microscope consisted of a frame, holding a glass tube connected to a lens holder into which various lenses, each mounted between brass plates, might be pushed. The simple microscope is essentially nothing but a brass plate encasing the lens, with

Fig. 5. Compound microscope made by Edmund Culpeper. Culpeper introduced the tripod stand ca. 1720, which was copied throughout Europe for half a century (Museum Boerhaave, Leiden).

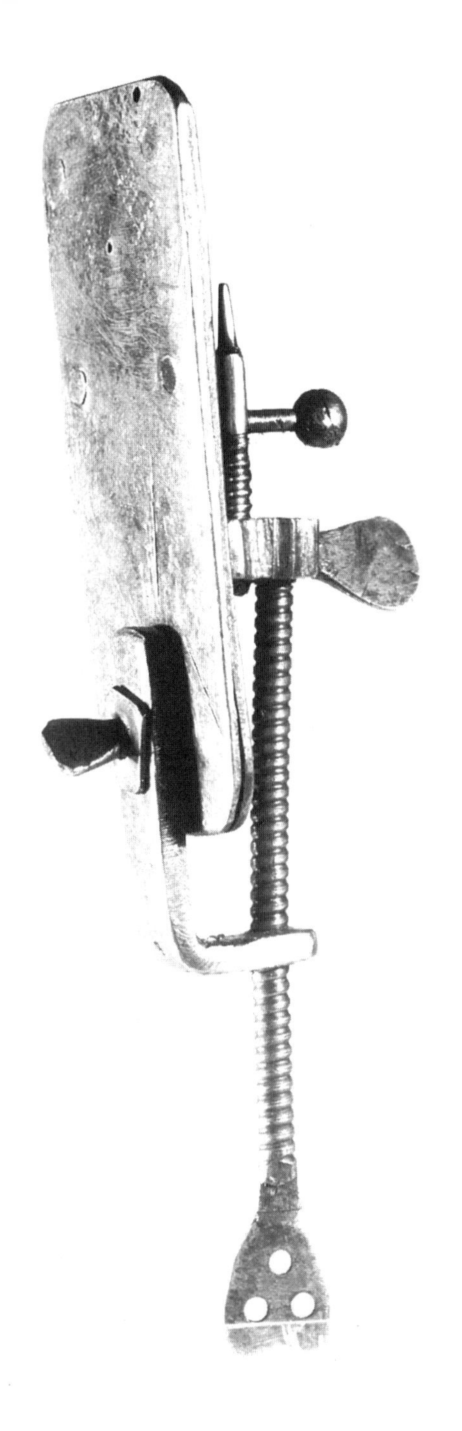

Fig. 6. One of Leeuwenhoek's microscopes (Museum Boerhaave, Leiden).

provisions for fixing the object in front of the lens, and a focusing mechanism. Whereas Leeuwenhoek's microscopes were of crude design and hard to handle, a variety of easily handled models could be acquired commercially by the beginning of the eighteenth century.

As a result of Leeuwenhoek's exciting discoveries (particularly of the spermatozoa), Huygens set out to design a simple microscope for his own use. He was assisted in this venture by his compatriot Nicolaas Hartsoeker. Their device was demonstrated to the members of the Académie des Sciences in Paris in the course of several of their meetings in 1678.[25] As soon as Huygens had constructed this instrument he embarked in earnest on his microscopic research. From June 1678 until early in 1679 a series of almost daily notes concerning microscopic studies is found in Huygens's manuscripts. In the course of these months Huygens sketched several alternative designs in his notebooks and in letters to his brother Constantijn.[26] The sketches for these alternative models are so intimately mingled with his notes on the animalculae as to show clearly that ideas for improvements occurred to Huygens in response to inconveniences perceived in the course of his investigations. From his sketches it appears that, essentially, the variant versions of Huygens's single-lens microscope consisted of a double frame of some eight to twelve centimeters in height. The front half of the frame contained the lens, and the back half accommodated a diaphragm. The twin parts of the frame were clamped together at one end, while the other end was held together with a screw. By turning the screw the distance between the two parts of the frame varied, and thus the specimen was brought into focus in front of the lens.

The design of Huygens's microscope contained two new features: a diaphragm, which he developed in order to exclude stray light from the object so as to improve the quality of the image, and a specimen revolver. These devices found ready acceptance and were copied by several instrument makers. Microscopes very similar to Huygens's design were put into production by the Paris instrument makers Michael Butterfield, Louis Chapotôt (or his son), and Depouilly, as well as by the Augsburg artisan Cosmus Conrad Cuno.[27]

In the Netherlands the production of simple microscopes was taken up

Fig. 7. A Musschenbroek microscope for low magnifications (Museum Boerhaave, Leiden).

by the instrument makers Musschenbroek. Samuel van Musschenbroek designed two versions of this instrument: one for low and another for high magnifications. Musschenbroek's microscope fitted with low-power lenses was of a very simple design: it consisted of a thin, pointed rod, fitted with a handle, its top carrying the lens, which was mounted in an eyecup and could easily be changed for another. The object holder, fitted to an arm, could be clamped at the lower end of the rod. This object holder could easily be moved in any direction and at any distance from the lens by means of the ball-and-socket joints or "Musschenbroek nuts" in the arm (see figure 7).

Their microscope intended for high-power lenses was of a more sophisticated design: the interchangeable lenses were contained in a lens

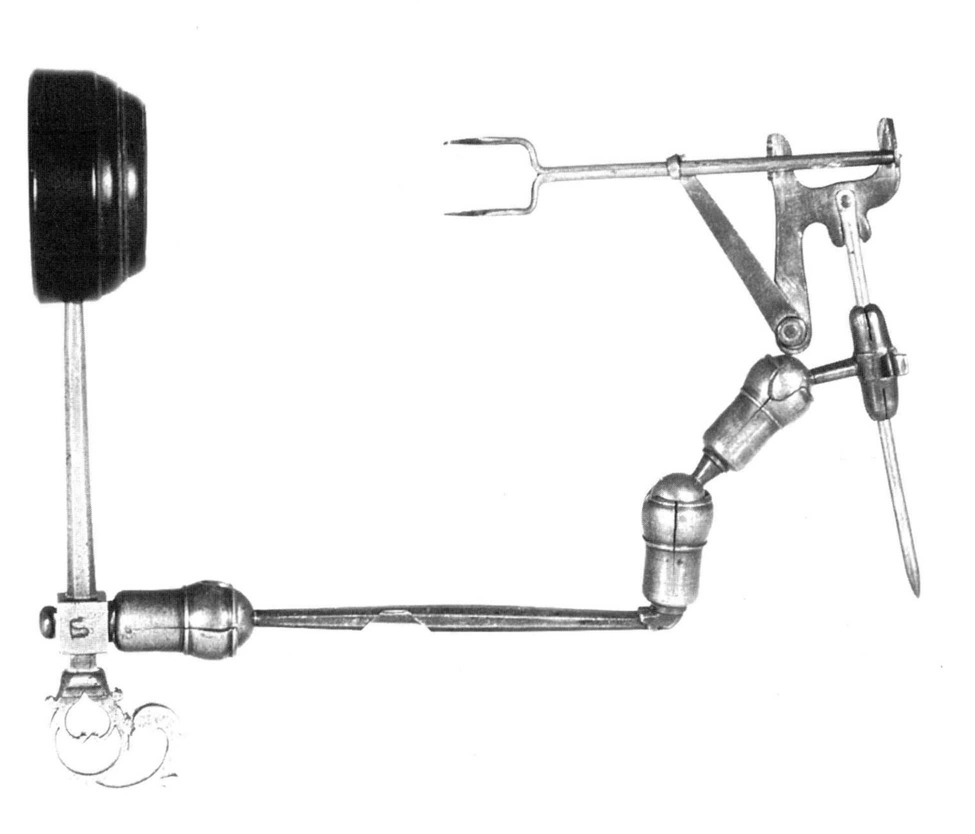

holder, which was fitted with a sector drilled through with diaphragms of different sizes. This lens holder was attached to a handle, at the lower end of which the specimen carrier was connected. Between the two legs of the instrument a spring was mounted. Focusing was effected by screws that moved the two legs of the instrument closer to, or farther away from, each other. This focusing mechanism was widely applied, particularly during the second half of the eighteenth century, in what is called the compass microscope.

The most successful design, from an entrepreneurial point of view, was no doubt the screw-barrel microscope, first proposed by Hartsoeker in 1694 in his *Essay de dioptrique*. After James Wilson, optical instrument maker in London, had advertised this model in the *Philosophical Transactions* of 1702, the screw-barrel microscope gained great popularity and was produced in large numbers by several instrument makers, Culpeper among them. This microscope was constructed from two tubes that screwed into each other, the outer tube containing the lens and the spring stage. The inner tube was used for focusing the object, wedged between the spring stage. The focusing tube was also fitted with a lens, the condenser, which ensured adequate illumination. Like the compound microscopes, most eighteenth-century simple microscopes were sold with an array of accessories.

Early Use of the Microscope

Although before 1660 few scientists thought of embarking on microscopic research, between 1620 and 1660 several people made a superficial inventory of the subjects that could successfully be investigated with the microscope. Records of a number of investigations were published as separate treatises, such as Odierna's *L'occhio della mosca* (1644), or as a chapter in more comprehensive books such as Fontana's *Novae coelestium terrestriumque rerum observationes* (1646).

The first object studied with the microscope, according to extant sources, was some kind of insect. In 1610 a friend of Galileo noted that he had "heard a few days back the Author himself [Galileo] narrate . . . amongst others in what manner he perfectly distinguished with his telescope the organs of motion and of the senses in the smaller animals, and especially in a certain insect in which each eye was covered with a rather thick membrane, which is perforated by seven holes like the visor of a warrior to allow it sight."[28] Although Galileo occasionally executed similar observations over the ensuing years,[29] no one seems to have followed suit until the Keplerian type of microscope was introduced in approximately 1620.

As mentioned earlier, one of the first people to acquire a compound microscope of this type was Peiresc, a patron and amateur of science. As his attention was easily caught by novelties in science, he seized the opportunity to become acquainted with the microscope when he met Kufler in 1622. Peiresc's account of his first experiences with the microscope, which has been preserved among his extant manuscripts, indicates the preferences in early microscopic observations and the kind of details that struck receptive minds like his. Peiresc relates how he and his companions saw several insects of which "the smallest seemed as well formed as the biggest."[30] He observed the eggs of fleas and lice, and in some type of louse he observed darkish structures, which he fancied were filled with blood or excrements. These observations made him "admire in the highest degree the effects of divine providence, which was far more incomprehensible to us when that aid to our eyes was wanting."[31] Peiresc, therefore, after having examined several kinds of arthropods, was struck by the numerous details revealed by the microscope. Moreover, he realized that microscopic investigation might lead to a deeper understanding of the smallest specimens of living beings.

Just as the earliest written records of microscopic investigation are concerned with insects or other arthropods, the oldest extant illustration (now very rare), dated 1625, depicts an insect, a bee. Some members of the Accademia dei Lincei—Prince Cesi, Francesco Stelluti, and Francesco Fontana—joined forces in a detailed study of the bee. Their combined efforts were published in 1625 in the form of two broadsheets: a printed one containing a natural history of the bee, entitled *Apiarium* and signed by Cesi, and an engraved one illustrating the magnified external parts of the bee, headed *Melissographia* and signed by Stelluti. A few years later, in 1630, Stelluti published a slightly different copy of the illustration of the bees in his edition of Persius's poems (see figure 1); that illustration was accompanied by a much more detailed description of the external parts of the bee. Stelluti noted about the eyes, for instance, that "of the three parts of the head, two are almost completely occupied by the eyes, which are very large and ovate, having the more pointed part on the lower side of the head, they are all hairy, and the hairs are placed like a checkerboard, or a gridiron, or a net, as they are on the eyes of all other insects that fly, which also resemble gridirons."[32]

The investigations of Galileo, Peiresc, and Stelluti were all concerned with the outer appearance of the insect body and a few other objects. In fact the majority of references to microscopic observations in the period prior to 1660 are concerned with the external details of the bodies of insects, and these organisms were to remain favorite objects for a long time to come.

From a variety of sources it appears that the microscope gradually became known as an observational aid among scientists living far apart and with widely diverging interests. Some, like Isaac Beeckman,[33] Sir Théodore Turquet de Mayerne,[34] and Sir Thomas Browne,[35] proved to be well informed about the microscope, even though it is not always clear whether this was from their own experience or from hearsay. Others, such as Francesco Fontana and Athanasius Kircher, included short inventories of recent microscopic investigations in their published works.[36] These inventories emphasized that the microscope added a new dimension to natural history, but the full potential of the instrument did not come to the fore.

However, a short treatise on the eye of the fly was published more or less simultaneously by Gioanbatista Odierna; this reports a momentous microscopic enquiry. Odierna, a priest by profession, lived nearly all his life in Sicily.[37] In his spare time he kept abreast of the latest developments in science, corresponding with some of the greatest scientists of his day and engaging in original research. Among other things, he observed the moons of Jupiter, studied the curious appearance of Saturn, kept meteorological records, and published the most advanced microscopic research of the period before 1660.

In his *L'occhio della mosca,* published in 1644, Odierna described how he studied the eye of the fly from the outside and dissected it to investigate its internal structure. He first established that the outside of the eye consisted of thousands of similar, square, somewhat bulging segments or facets (see figure 2). Having opened the eye, he then found the eye to be constructed from four separate elements. From outside to inside these were (1) a solid transparent layer, shaped into facets, (2) a layer of pyramidal bodies, (3) a dark membrane on which the apex of the pyramids converged, and (4) a small particle of brainlike material, enclosed within the dark membrane, which was not connected with the brain.

On the basis of these findings Odierna concluded first, that the insect eye is composed of thousands of identical units (the pyramidal bodies and their corresponding facets), each being a veritable eye. His second conclusion was that each eye could only pass on to the retina (the dark membrane) light rays falling perpendicular to the surface of its facet. His third conclusion, and the one he deemed to be the most important, implied that light perception occurs in the small particle inside the retina. Therefore according to Odierna the insect eye both receives and perceives the multitudinous images of the outside world, and, consequently, visual perception in insects occurs outside the brain.

Whereas preceding investigators had observed primarily the exterior of their objects, Odierna endeavored to reveal the construction of the interior of the insect eye. In order to do so the eye first had to be boiled and dried in the sun, and Odierna advised that "the experienced anatomist cuts with the extremity of the sharpest possible knife (prepared for this task), while his eye, brought very near to the upper opening of the instrument [microscope], watches acutely, cuts the surface of the cornea, in order to separate it from the remaining substance."[38]

The combination of refined anatomical procedures with microscopic observation forms one novel aspect of Odierna's study. A second one is that he used his knowledge of the anatomical structure to explain the action of the fly's eye. By this time, the middle of the seventeenth century, the scientific elite appreciated the microscope and its potential usefulness. More and more writers referred to recent microscopic discoveries in their publications; among these were Walter Charleton,[39] Joseph Glanvill,[40] and Domenico Panarolo.[41] Extant manuscripts reflect the mounting interest in microscopic research among people on the periphery of science: witness the manuscripts of Peter Mundy, a traveler, who became acquainted with the microscope ca. 1647 and left posterity a drawing of the head of a fly.[42]

In 1656 Pierre Borel published a collection of a hundred miscellaneous microscopic observations as an appendix to his enquiry concerning the true inventor of the telescope, entitled *De vero telescopii inventore* (1655).[43] In his *Observationum microscopicarum centuria* Borel described his investigations of plants, animals (mostly insects), human material, and minerals. Borel made few contributions as an observer. He saw, among other things, the compound eyes of insects, the beating of the heart in a spider, and the flow of the blood in a louse; and he investigated the texture of the human heart, liver, kidney, and testicles. He described what he saw very briefly and in no great detail, and the illustrations he added are very crude (see figure 3). It is thus difficult, if not impossible, to determine what kinds of organisms he actually saw. However, the importance of his microscopic treatise lies not so much in the content, but rather in the direction taken by his speculations.

Just as for some of his predecessors, the aspect of the complex organization of the insect body reinforced Borel's admiration for the Creator of all living beings. He felt that such spectacles ought to convince atheists of the existence of God. Indeed, he wrote, as the microscope makes

> atoms as it were visible, and small insects are transformed into gigantic monsters, with its help innumerable parts are discerned in those living atoms and daily the doors of a new physics are opened, to such an extent that God's Majesty comes to light more clearly through these small bodies than through gigantic ones, and their stupendous construction convinces even the most atheistic and leads them to awareness, admiration and veneration of their Supreme Maker.[44]

Borel was not content just to observe the appearance of his objects, he also manipulated them so as to reveal their structure. Thus he found the insect body to be furnished with "feet, nerves, eyes and all animal parts."[45] More important, Borel extended the scope of microscopic research to the human body, speculating on the application of the microscope in medical practice and assuming that, at the microscopic level, the physiological processes were brought about by mechanical action. Borel investigated the blood, skin, and hair of the human body and the texture of various organs. He also investigated some pathological states such as ingrowing eyelashes and the discharge from ulcers. Where the employment of the microscope in medicine was concerned, Borel entertained the idea that in certain cases material signs of imminent death may be detected microscopically.[46]

As his references to "atoms" in the earlier quotation have already indicated, Borel was very much conversant with the new emphases in contemporary science. He also relied, however, on the principles of mechanical philosophy when reflecting on the use of the structures he had observed, writing, for instance, "You will find the heart, the kidneys, the testicles, the liver, the lung and other parenchymatous organs to be plaited from structures and fibres, a sieve as it were, through which various substances are separated by nature, according to the shape of the holes, thus passage is given only to certain atoms of a certain shape."[47]

Even though very few of the publications wholly or partly concerned with microscopy were published between 1620 and 1660, there was fitful interest in the subject while its possibilities were being explored. The various endeavors in this early period of microscopy culminated in Odierna's and Borel's contributions. Odierna's study of the fly's eye demonstrated convincingly that the prospect of the microscope disclosing the secrets of organic matter could not be realized until the object was taken apart to reveal its structure. His skillful union of the optical tool and anatomical technique indicated an appropriate method for further exploration of the structure of living material.

Both Odierna and Borel attempted to support their findings with an explanation of their operation, which was rooted in mechanical philosophy. Indeed, the refraction of light and "atoms of a certain shape" are topics that feature prominently in that philosophy. Both men were very much immersed in the contemporary development of science, and a wide range of subjects, from the physical sciences as well as the life sciences, attracted their attention. Therefore, their mechanical explanations of physiological processes conform to the general direction of these two investigators' speculations. Since Odierna's *L'occhio della mosca* and Borel's *Observationum microscopicarum centuria* are both indisputably highlights of microscopy before 1660, it is clear that microscopic observation only came to fruition once it gained support from microtechnique, however primitive, and an explanatory schema.

The influence of Odierna's and Borel's two publications on their contemporaries and immediate successors is hard to assess. Few direct references are made to these two booklets in subsequent publications, but as the one was concerned with a single very specific subject, while the other covered a hundred different subjects very superficially, this is not altogether surprising. However, both books were certainly appreciated by workers in the same field. Henry Power, for instance, modeled the microscopic part of his *Experimental Philosophy* on Borel's treatise, and he longed to see a copy of Odierna's.[48] Early surveys of the history of microscopy, such as Sachs's and Buonanni's, also took note of these treatises.[49] Yet any statement to the effect that Odierna or Borel headed microscopy in a new direction certainly overvalues their influence. Nevertheless, their researches affirmed the relevance of microscopic investigation with regard to the elucidation of the structure and operation of organic matter.

It appears from both Odierna's and Borel's writings that reports of microscopic research were from the very beginning imbued with theological implications. The hand of God was recognized in the intricate construction of the smallest living beings. This theme was to reach its apotheosis in eighteenth-century physico-theology, which derived a number of arguments (and not the least relevant) from microscopic observations.

Establishment of Microscopic Data

Whereas prior to 1660 microscopic observations frequently appear to have been undertaken in order that scholars might marvel at the wondrous sight of hugely magnified crawling vermin, the nature of such investigations changed during the 1660s to the purposeful study of organic structure. The resulting determined application of the microscope caused the concomitant development of the basic working methods of microscopy. Microscopists, their critics, and physicists at the same time discussed the fallacies inherent in the use of the microscope. Four aspects relevant to the establishment of microscopic data will be discussed here: the preparation of the object, an estimation of the size of the observed structures, the interpretation of the image with regard to optical imperfections of the devices employed, and communication of the observations to the scientific community.

Although the earliest microscopists wrote in glowing terms about the wonders to be observed on "flies the size of an elephant," it gradually became obvious that objects other than minute insects do not yield much information when simply placed beneath the microscope. The investigator must take his object apart in an orderly fashion and also take measures to increase the distinction between the various structures.

The critical importance of deliberately taking the object apart so as to get to know its structure is perhaps best illustrated with Hooke's discovery of the pores or cells in cork (figure 8). Hooke started his investigation of cork with a rather thick piece, the surface of which he had smoothed. He thought he perceived cork to be porous but could not distinguish the details very clearly, but, after he had cut an exceedingly thin slice off the piece of cork, he was able to distinguish very plainly that cork was indeed porous.[50] Subsequently Hooke also investigated a piece of cork cut at right angles from the first piece and in so doing discovered that the pores "were not very deep, but consisted of a great many little Boxes, separated out of one continued long pore, by certain *Diaphragms*." Thus by methodically sectioning a piece of plant material Hooke discovered the confines of what later became known as the plant cell.

Subsequent investigators of plant structure benefited from Hooke's experiences. Grew's investigations of the smaller details in the anatomy of plants, for instance, were conducted along the lines indicated by Hooke. Indeed, Grew systematically studied very thin slices of plant material cut at different angles from the root, stem, leaves, or other part of the plant,

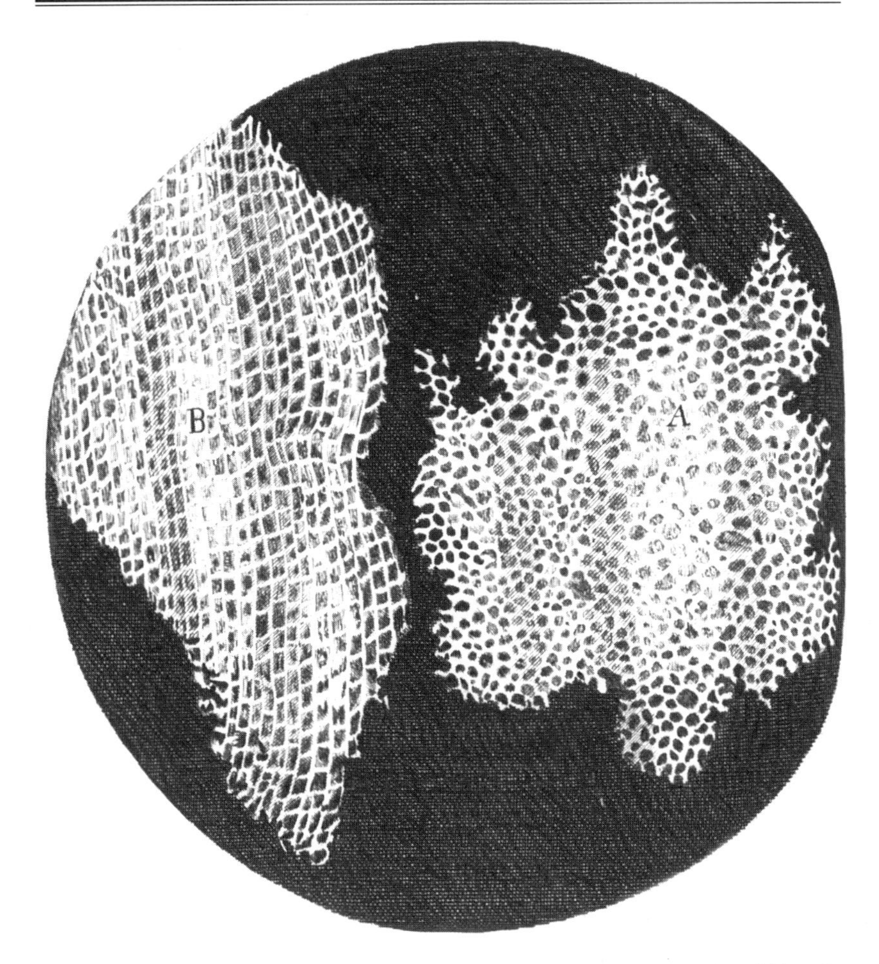

Fig. 8. Hooke's illustration of the cells in cork as seen in a cross section (A) and longitudinal section (B), table 11 in his *Micrographia* (1665).

being well aware that only a methodical approach would serve his ends. Plant anatomy must be studied, he wrote, "By several Ways of Section; Oblique, Perpendicular, and Transverse; all three being requisite, if not to Observe, yet the better to Comprehend."[51] Thus the spatial arrangement of the structures of which vegetable objects are composed proved relatively easy to fathom by judiciously relating one section to another.

The texture of animal tissues on the other hand could not be penetrated so easily, because the softness of most of that material prevented the cutting of thin slices. However, as microscopic investigation of animal materials was pursued within the context of refined anatomical research or *anatomia subtilis,* several of the techniques employed in macroscopic anatomy were also used in combination with microscopy. A number of artifices, such as boiling, maceration, and injection with various substances, were used to elucidate the texture of organs such as the liver, kidney, brain, and skin. These methods served to make the component parts of the organs more accessible for inspection. Malpighi's inquiries into the fabric of the tongue and kidney are cases in point. Malpighi first boiled the tongue so that he could easily peel off the two outer dermal layers separately, exposing the papillary body underneath.[52] The details of the construction of the various layers could thus be inspected, and Malpighi subsequently proposed a mechanism for taste on the basis of these observations.

The injection of materials such as wax, mercury, or gypsum into the vessels of the animal body was developed as a technique of anatomical research in approximately 1660.[53] Although usually employed for studying the distribution of vessels of macroscopic size, some microscopists made good use of this technique as well. Malpighi, for instance, injected the arteries of a kidney with ink prior to microscopic inspection.[54] As a result of this procedure the glomeruli showed up as black globules, the first support of Malpighi's theory of a basic glandular unit.

Coloring agents as a means of increasing the contrast between various structures were used infrequently.[55] Leeuwenhoek once used saffron as a means to color muscle tissue with the express intention of making the structures more clearly visible for his draughtsman.[56] The application of a coloring substance to animal organs, or use of other artifices, revealed a number of details of the organs' structure that in their turn suggested new concepts concerning these organs. However, some of the techniques employed damaged the structures present or caused artifacts, thus giving rise to errant interpretations. Willem Weyer Muys, for instance, observed that the colored fluid he injected into the blood vessels of a piece of muscle also colored the muscle fibers themselves.[57] On the basis of this observation Muys concluded that the muscle fibers are hollow and have an open connection with the blood vessels. Malpighi's coloring of the boiled brain was of greater consequence for contemporary physiological thought, as it led to a very definite view on the part of the investigator concerning the microscopic anatomy and functioning of the brain.[58] Malpighi colored a piece of his specimen by dropping ink on the surface, and after waiting a few minutes for the ink to be absorbed, he then wiped off the surplus. Malpighi found noncolored patches, which he took to be glands, nicely outlined in the brain tissue. It has been conclusively demonstrated that Malpighi in this case based his views on an artifact.[59]

Even though Malpighi did not consider whether his manipulations might not have damaged the original structures he examined, some investigators suspected that the various treatments of the preparation, the fact of removal of the specimen from the body, or the desiccation of the object during inspection might be the cause of irrevocable changes in the material. Swammerdam, for instance, having seen the blood globules in a sample of blood, doubted whether it also has globules when still enclosed in the vessels in a fluid state.[60] He proposed to investigate this by inserting a glass tube into an artery in a live dog, and observing the blood flowing into the tube through a microscope. Swammerdam did not perform this particular experiment, nor did anyone else. However, Leeuwenhoek in the course of his various investigations concerning the circulation of the blood frequently saw the blood globules in vivo. His own reactions to his observations during the several stages of this study demonstrate the persuasive force of observations on the living organism. Although Leeuwenhoek had observed the capillaries of the blood in many specimens, he was not wholly convinced that these constituted the final link between arteries and veins until he saw the propulsion of the blood globules through the capillaries in the gills of a tadpole, simultaneously with the heartbeat of the animal. This spectacle he deemed the most exciting of his entire research up to that time.[61]

Several investigators realized that changes in the object were caused by desiccation. Leeuwenhoek, for instance, observed a number of holes in a section of the optic nerve of a cow.[62] These holes, he realized, were generated by the evaporation of volatile particles from his preparation. Desiccation could be prevented by viewing the object covered with a tiny amount of a clear liquid—water or oil—a procedure that had the added advantage of showing detail more clearly. Hooke found that in this manner an admirable fabric could be discovered in many animal and vegetable bodies.[63] Even though desiccation might be the cause of deleterious effects in one's preparations, dried specimens could be preserved for any length of time, which might prove useful. Leeuwenhoek, in particular, kept dried and mounted specimens for a long time, and occasionally used these for supplementary investigations.[64]

The techniques employed by the various microscopists were often developed to meet the needs of specific enquiries. Malpighi, when studying the development of the chicken embryo, was confronted with an indistinct view of the structures as he had to observe the embryo in situ on the yolk of the egg. This involved putting the opened, but otherwise intact, bulk of the egg underneath the objective of the microscope, causing obvious difficulties. In order to get a clearer view of the embryo Malpighi developed a method for removing the entire blastoderm from the yolk, and examining it spread out on a glass slide.[65]

These techniques, therefore, were wholly individual and seldom explicitly described, even jealously guarded on occasion. Hooke, who experienced great trouble in devising a microscope and method for observing the animalculae in pepper water (partly because Leeuwenhoek did "not think fit to impart" his methods), published, in reaction to Leeuwenhoek's secretiveness, a short essay in 1678 entitled *Microscopium* in which he described a variety of methods of his own invention. One of these was the use of very small mica or glass plates for covering objects with irregular surfaces so that misleading optical effects might be avoided. Simultaneously, Christiaan Huygens also employed mica coverslips for covering samples of infusions, a procedure that he, too, thought of as an original invention.[66]

Apart from Hooke's essay no surveys of microtechnique were published until the mid–eighteenth century, when introductions to microscopic science for the layman became popular. The instructions for the preparation of one's objects in these books were largely concerned with methods of mounting objects on or in suitable containers, such as glass slides, live boxes, glass tubes, or capillaries.[67] The emphasis was on the long-term preservation of an object, rather than on the amelioration of detail and resolution of structures.

The dimensions of an object or its component parts either may be computed by comparing the specimen to an object of known size, or can be measured directly with a suitable measuring device. Both Hooke and Leeuwenhoek used a comparative method. Hooke placed a ruler marked in subdivisions of an inch next to his microscope and managed to look simultaneously with one eye through the microscope at his object and with the other eye at the ruler "to cast, as it were, the magnifi'd appearance of the Object on the Ruler, and thereby exactly to measure the Diameter it appears of through the Glass."[68]

Since he also measured the size of the object directly with the same ruler, he could estimate both the magnification obtained and the dimensions of structures observed in the objects. Leeuwenhoek's method was based on estimating the size of an entire object or a certain structure in relation to a standard object, such as a grain of sand. Assuming that eighty sand grains lined up would cover an inch, he could then compute the dimension of the object or structure, or calculate the total number of these objects that could be contained in a given space. Leeuwenhoek initially took the size of a grain of sand as a starting point for his computations; in later years he changed his standard to the diameter of a hair.[69] Other investigators also used natural standards. Muys, for instance, used the diameter of a red blood cell as a standard for measuring the diameter of the various categories of muscle fibers.

At the beginning of the eighteenth century several methods for direct measurement were developed. In 1710 Theodor Balthasar described in his *Micrometria* how the principle of the screw micrometer, which had been applied since the middle of the seventeenth century in telescopes, could also be employed in microscopes.[70] In the *Philosophical Transactions* of 1718 James Jurin described another method for direct measurement.[71] He twisted a silver wire tightly around a pin in such a way that no spaces were left between the coils. When he had thus covered a stretch of one inch he simply counted the number of coils and computed the diameter of the wire. He then cut the wire into tiny pieces and scattered these on the object plate. Thus he could observe the pieces of wire simultaneously with the object and thereby measure the object.

However, it was not until the mid–eighteenth century that various kinds of micrometers, some based on either of these two methods, became commercially available.[72] Up to that time an estimate of the dimensions of an object was only infrequently attempted. In fact Leeuwenhoek had a predilection for marveling at the minuteness of the texture of living matter, one he supported by large numbers of presentations of objects of extremely small dimensions.

The image rendered by seventeenth-century microscopes was subject to serious flaws such as indistinctness, limited fields, and colored fringes surrounding the object. These flaws were universally recognized and accounted for during observation and interpretation. Hooke warned in the preface to his *Micrographia* that although the microscopes will greatly magnify an object, the apertures of the object glasses are so very small that the object appears dark and indistinct.[73]

It has been established experimentally that the optical image of seventeenth-century microscopes suffered to such an extent from various optical defects that illusory images were observed by some microscopists, for example, Athanasius Kircher's famous reference to the worms in the blood of feverish people.[74]

Theodoor Kerckring is probably the best-known contemporary critic of microscopic observation. He warned against uncritical acceptance of microscopic images. "You know," he wrote, addressing himself in particular to Malpighi, "that the range of vision is small. You know that the colors vary so much that by observation by these means, something is put into things so that it is difficult to distinguish what is original and real. You know also that it is possible that things seen in this way seem to be apart while in reality they are continuous."[75]

Chérubin d'Orleans stressed that, when small creatures are examined, only part of the object can be seen at a time, so that the investigator has to join the several images as best he can into one, to obtain a picture of the whole of the creature. He felt sure that some readers would find these images "very changed, disproportionate, and differmed with additions and subtractions completely conjectural."[76] However, since Chérubin completed this passage with the recommendation of a binocular microscope of his own invention, which purported to overcome these particular difficulties, the alleged flaws of other microscopes are no doubt exaggerated, yet worthy of consideration. This kind of problem could be countered by a prudently executed exploration of the object. Malpighi, in response to Kerckring's admonitions, stated the principle clearly:

> If we use microscopes in the proper way, no deception whatever arises for the parts should be explored first with one having a single, less acute lens, with which a larger mass is visible and continuity and discontinuity are distinguished, and then a more acute lens should be employed. This enlarges a very small mass, and when it is gradually moved, certainly reveals to the eyes the parts of the object as continuous or separate, just as we see the letters when the eyes are moved slowly in reading.[77]

Yet the true appearance of the structures observed remained elusive. Hooke often found it difficult to discover the true shape of the things he studied because the same object seemed quite different in one position to the light than in another. As an example he mentioned the compound eye of insects: "The Eyes of a Fly in one kind of light appear almost like a Lattice, drill'd through with abundance of small holes; . . . In the Sunshine they look like a Surface cover'd with golden Nails; in another posture, like a Surface cover'd with Pyramids; in another with Cones; and in other postures of quite other shapes."[78]

Nevertheless the microscopists, through painstaking research and careful interpretation, arrived at definite views as to the structures to be observed in their objects. Even if some doubted the veracity of certain particulars,[79] an abundance of microscopic data testified to the fact that the microscope magnified previously invisible small structures to readily perceived dimensions. Even so, it did not increase the acuity of normal vision. This point was made by Chérubin d'Orleans in his account of someone's all too enthusiastic description of some minute insect. He wrote that if you see an insect through the microscope enlarged to the size of a grain of salt of which you can see the details with the bare eye, it is then not astonishing to discover many details on the body of this insect. However, if this magnified insect appears no bigger than a very small grain of salt of which no details can be seen, then you cannot detect the details of the body of this insect.[80]

Investigators could communicate their microscopic data to the scientific community either verbally or pictorially. In fact, they often combined descriptions and illustrations into one discourse. Since the early microscopists were disclosing a hitherto hidden world, a terminology for the various structures had yet to be developed, as well as a convention for microscopic illustration.

A variety of terms was used to denote the same structures. For instance, the words "cell," "pore," "bladder," "utricle," and "globule" were used indiscriminately in reference to what is now defined as the plant cell. Conversely, the same term could be applied to a wide range of structures. The word "globule," for instance, was not only used by at least one microscopist to denominate the plant cell, but at the same time was used by the same writer to indicate the red blood cell as well as drops of fat. However, the microscopists could express themselves well enough to be able to convey to their readers a mental picture of what they wished to communicate. Goverd Bidloo was thus able to include a picture of the texture of the brain cortex in his anatomical atlas of 1685, which was clearly based on Malpighi's description and not on personal observation.

However carefully and precisely a description might be worded, a drawing is obviously more illustrative and easier to grasp. Glancing through microscopic illustrations from the seventeenth century, it strikes one that numerous drawings of the anatomy of insects were produced, and that the microscopic structure of plants was abundantly illustrated. The microscopic anatomy of the animal and human body, on the other hand, was poorly illustrated by those investigators who studied those structures most closely. This discrepancy in the number of pictorial representations of materials of different origins is a consequence of the different qualities of these materials. These determined both the methods suited to elucidating the various objects, and their intelligibility. Most plant material is easily sectioned and its structure stands out clearly. The minute parts composing the insect body may be separated with relative ease and distinguished through a slight contrast between them. The substance of human and animal organs, on the other hand, was not easily sectioned, and the contrast between the various component parts is very poor when uncolored.

The details of the object, once clarified to the satisfaction of the investigator, had to be translated into a drawing. The delineation of vegetable material is facilitated by the clear patterns observed in sections, and a fairly long-standing tradition in anatomical illustration provided some support in depicting insect anatomy. The presentation of animal and human microscopic anatomy had the most need of a gifted draughtsman, particularly since as a rule lumps of material were investigated, and therefore three-dimensional structures had to be represented. The illustrations of microscopic anatomy in Bidloo's atlas demonstrate that the general impression of the microscopic appearance of such a lump could be recorded satisfactorily, but that a convention to illustrate precise anatomical details was lacking.[81] Bidloo, or perhaps his draughtsman Gerard de Lairesse, developed such a convention in a few instances, as for instance in the case of the stomach wall (figure 9). By depicting from right to left the aspect of the three successive layers of the stomach wall they endeavored to represent its three-dimensional microscopic structure.

Hooke carefully indicated the order of magnification of the objects depicted in his *Micrographia,* in order that the readers might gain a clear idea of the actual size of the objects, and also be able to discern these particulars in their own preparations.[82] By the same token Leeuwenhoek often calculated and reported the dimensions of the parts he described, and other investigators, Swammerdam for instance, tried to provide an indication of the size of the particulars illustrated by depicting both the object as seen by the naked eye and as seen through the microscope. There was, however, no agreed method by which to indicate the actual size of the structures detected by the microscopists. This hampered the growth of a body of shared, entirely corresponding knowledge about microstructures among the learned.

From the above it appears that men such as Malpighi, Swammerdam, Hooke, Leeuwenhoek, and Grew devoted the best of their attention, skill, and ingenuity to the preparation of their objects. Each man had to invent ways and means to draw the secrets from his specimens *de novo.* The na-

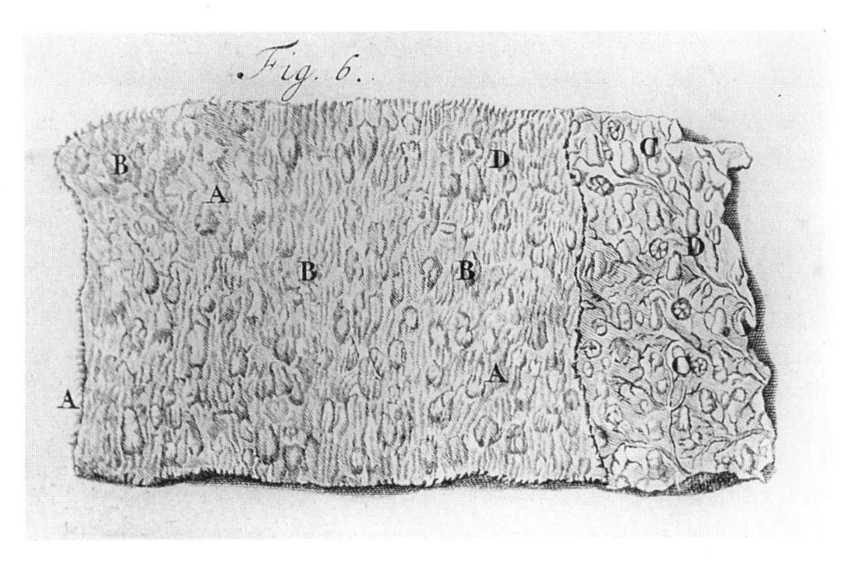

Fig. 9. Bidloo's representation of the wall of the stomach, in which several superposed layers are depicted, from table 35 in his *Anatomia humani corporis* (1685).

ture of their exertions, and rewards, were most eloquently voiced by Malpighi, who wrote

> My dissertation on *Bombyx* was an occupation to the last degree laborious and fatiguing, because of the novelty, minuteness, fragility and entanglement of the parts. Hence the prosecution of the task made it necessary to develop entirely new methods. And since I pursued this exacting work for many months without respite, I was afflicted in the following autumn with fevers and inflammation of the eyes. Nevertheless in accomplishing these researches, which brought to my notice so many strange marvels of nature, I experienced a pleasure which no pen can describe.[83]

Even though individual microscopists took great pains to glean the least detail from their materials, they usually only referred to the methods they employed in a cursory manner. Moreover, the microscopists of this period, with the exception of Leeuwenhoek, made only temporary specimens for immediate personal use, which encouraged ad hoc solutions to the problem at hand. Consequently, a generally current microtechnique could not be developed, and a critical assessment of the merits and disadvantages of the various methods was prevented.

Whereas the effects of manipulations of the object with regard to the latter's effectiveness or damage to the structure were not a subject of discussion in contemporary scientific literature, the defects of the optical image were widely discussed. Consequently the most striking aberrations in the optical system of the microscope, as well as some of the fallacies of the image that thwarted interpretation, became widely known. Nevertheless, discussions concerning the trustworthiness of microscopic images were largely marginal to the contemporary appraisal of microscopic discoveries, and there was no serious doubt that the microscope offered a reliable view of the fabric of organic structure.

Microscopic Publications

A memorable event marking the onset of the first heyday of microscopy was the publication of Hooke's celebrated *Micrographia* in 1665. Even if it was not the first, nor even the most important, publication in seventeenth-century microscopy, it was certainly the most persuasive and impressive survey of the microworld to be published in that period.[84] Numerous other books deserve equal fame, to name but a few examples: Malpighi's *Dissertatio epistolica de bombyce* (1669), Swammerdam's *Ephemeri vita* (1675), and Grew's *The anatomy of plants* (1683).

For the purpose of the present enquiry into the aims, pursuits, and failures of microscopic investigations it is opportune to demarcate the first flourishing of the use of the microscope with some precision, and the bibliometric method may be used. In this case the publications concerned with microscopy that appeared during the seventeenth and early eighteenth century have been listed, and the total quantity of microscopic research published has been calculated. The aim is not to present a thorough bibliographical analysis of microscopic science during this period but rather to measure, with readily available means, the scientific output of microscopy during the relevant period. The underlying premise is that microscopists communicated their findings to the scientific community through some sort of printed matter. A quantitative appraisal of such communications will provide a reasonably accurate idea of the pace of the progress of microscopy in the seventeenth century.

Besides publishing their results in a book like the ones mentioned above, many investigators chose to divulge their data and thoughts via a contribution to a scientific journal. The use of the scientific journal as a means of communication between scholars was established in the early 1660s and was closely associated with the emergence of scientific societies in the seventeenth century.[85] Such journals were founded in virtually all the European countries during the second half of the century; some of these appeared only for a few years, while others have continued up to the present time. For the present purpose three leading journals from three different cultural spheres in Europe have been selected. These are the *Journal des sçavans* representing French intellectual circles, in which the Paris académiciens played the dominant part; the *Philosophical Transactions* as representative of the English virtuosi and their intellectual friends gathered in the Royal Society; and the *Miscellanea curiosa medico-physica* as representative of the German-speaking intellectual elite of middle Europe united in the Collegium Naturae Curiosorum. Each of these journals began its activities very early, in 1665, 1666, and 1670, respectively, and continued to appear throughout the period under consideration here.

The publishing policies of these journals were very different, however. The *Journal des sçavans* was conceived by its founder and first editor, Denis de Sallo, as a weekly journal, containing "useful information" on recently published books, obituaries, news of experiments in physics and chemistry, discoveries in the arts and science, meteorological data, and legal and ecclesiastical judgments. In short, it was intended "to transmit to readers all current events worthy of the curiosity of men."[86] The *Journal des sçavans* therefore concentrated on communicating news from all spheres of intellectual activity. To a large extent the editors selected their news from previously printed sources, and as a result this journal contained very few original scientific contributions. In fact, the scientific output of the Académie des Sciences was mainly published through another periodical, the *Histoire et mémoires.* However, this journal did not appear before 1702 and therefore only covers the second half of the period relevant to the present enquiry.

The *Philosophical Transactions* was a very different journal. From the very first issue the *Transactions* concentrated on reporting the scientific activities of the members of the Royal Society, as well as accounts of such activities sent to the members by their correspondents. Although each issue also contained abstracts and discussions of recently published books, these were never numerous enough to dominate an issue completely, as was the case with the *Journal des sçavans.*

As with the *Philosophical Transactions,* the policy of the *Miscellanea curiosa* ensured that reports of scientific activities formed the main component of each issue. The Collegium Naturae Curiosorum encouraged its members to undertake the study of some particular subject in the life sciences,[87] which could then be reported in their journal. The journal was published only once a year, and each issue included book reviews and obituaries in addition to numerous scientific reports, which were often very short. As the membership of the Collegium Naturae Curiosorum largely derived from the medical profession, the contributions published in this journal were often of a medical nature.

With the help of various sources, the publications concerned with microscopy between 1625 and 1750 have been assembled.[88] The beginning date of the period under consideration was obviously dictated by the very first publication containing a microscopic study, namely, Federico Cesi's and Francesco Stelluti's description of the bee in 1625. The ending date, however, was chosen arbitrarily as a convenient point in time not too long after the last of the prominent microscopists, Antoni van Leeuwenhoek, had died.

The assembling of a list of publications concerned with microscopy presents many difficulties. The question of how to determine whether a publication is concerned with microscopy is one of the first problems. Books such as Hooke's *Micrographia,* or Antoni van Leeuwenhoek's collected letters, naturally offer no problem because the microscope played such a dominant part in obtaining the data presented in these studies. Yet in many others the matter is not so clear. For instance, Malpighi's *De pulmonibus* of 1661 is famed as a milestone in microscopy, because in this letter to his mentor Giovanni Battista Borelli, Malpighi described for the first time the capillaries of the blood system within the lungs. Yet his elucidation of this particular feature of the lung's texture was only one among other, equally important discoveries. Moreover, he owed these discoveries less to microscopic inspection than to other investigative techniques, such as comparative anatomy and pathological observation. Yet in this publication Malpighi's final insights that put (as it were) the finishing touches to his investigation into the structure of the lung were gained with the help of the microscope. This particular publication is therefore included here as one concerned with microscopic research.

In other books the microscopic contribution toward the material presented is so slight that it must be ignored. For instance, Thomas Willis in his *Cerebri anatome* of 1664 referred briefly to his microscopic observations of the tissues of the brain and the nerves.[89] This research, however, was marginal to the main content of his book and the execution was superficial, yielding no results beyond those particulars observed by several contemporary investigators. Books like these are therefore judged as being unconcerned with, or irrelevant to, microscopic science.

A second problem is the question whether not only the first, but also all subsequent, editions of the relevant publications should be included in this inventory of microscopic literature. The assumption on which this bibliometric analysis rests is that the growth of microscopic science engendered a host of observations, data, ideas, and theories that investigators wished to communicate to people interested in science. Therefore, the first issue of a book or paper reporting particular observations or studies is directly related to the output of scholars concerned with microscopy. The publication of translations, second and later editions, and inclusion in opera omnia are more an indication of the continuing general interest on the part of the learned public in this subject and in the prominence of a certain author. Since this study is mainly concerned with the roots of microscopy, and therefore with the concerns, goals, and theories of those scientists actively involved with microscopy, it was decided to include only the first category in this list. In the case of Leeuwenhoek, therefore, a rather large number of his collected letters have been rejected, as these contain letters published earlier in the *Philosophical Transactions.*

In accordance with these criteria, popular introductions to the microscope, which began to be published toward the end of the chosen period, must be omitted as well. These books, such as Henry Baker's *The microscope made easy,* usually included—for the benefit of the "lovers of microscopy"—a survey of interesting subjects to be tackled, which were usually transcribed from works by the microscopists of the preceding decades.

The posthumously published works of seventeenth-century scholars presented a third problem. Swammerdam's voluminous manuscripts, mostly concerned with the anatomy and habits of various kinds of insects, were not published until a half century after his death. In our own time, on the occasion of the publication of his collected works, it appeared that Christiaan Huygens had also studied biological specimens through the microscope. What should be done in cases such as these?

As stated, the main idea of the present undertaking is to measure output by means of publications. From Swammerdam's correspondence with his friend Melchisedec Thévenot[90] it appears that Swammerdam had completed the manuscript for his *Biblia naturae* and had already prepared its publication when he died during the first weeks of 1680. Although Swammerdam had arranged in his will for its publication, the book was not published until 1737. Nevertheless, because it was obviously intended for immediate publication, for the present purpose this book is considered as having been published in the first years of the 1680s. Huygens's manuscripts are, however, an altogether different matter. There is no indication at all that Huygens ever contemplated the publication of his research on the infusoria. Indeed, in the manuscript of his *Dioptrica,* which was published posthumously in 1703, he gave only a general impression of the kind of work that might successfully be tackled with the microscope. Huygens's microscopic research is therefore not included in this inventory. In sum, included in the list are those publications (1) that were exclusively or largely based on microscopic research (such as Hooke's *Micrographia*), or (2) that contributed in a material way to microscopic science, even though a large part of the paper or book might be devoted to macroscopic observations (such as Malpighi's *De pulmonibus*).

Having listed the various books and articles (see appendices 1, 2, 3, and 4), the number of publications per five-year period, beginning in 1625, have been calculated. Before diagraming the total number of publications in chart 1 the two categories of publications—books and articles—have been weighted against one another. A monograph obviously bears more weight than an article in a journal, but how much more is an arbitrary decision. A book contains usually many more pages than an article. On the other hand, in an article the relevant data are usually communicated in a more succinct way, without the preambles and paraphernalia of a book. For the preparation of the chart it was decided to allot one point to books for every twenty-five printed pages, so that a book of 157 printed pages is given six points, but only one point is allotted to each article, whether it be large or small. Even before the inventory of books and articles was prepared, it was already obvious that Antoni van Leeuwenhoek contributed an enormous number of papers to microscopic literature. Leeuwenhoek's contribution to the total number of pages published has therefore been plotted separately.

Looking through the lists of articles devoted to microscopy in the selected journals, some interesting facts may be noted. The *Philosophical Transactions* contain by far the largest number of microscopic publications, which is in accordance with its publication policy. However, the published contributions are mainly those submitted by Leeuwenhoek, whose entire scientific output was largely published through this journal. Another interesting fact is that, although the *Journal des sçavans* contained very few original microscopic publications, most of these were, curiously enough, concerned with the compound eyes of insects. Finally, the *Miscellanea curiosa* published a fair number of microscopic papers, including many on insects that were mainly contributed by Johannes von Muralt, but also a relatively large number on pathological subjects, a topic that attracted barely any attention in the *Philosophical Transactions.*

From the chart it appears that production of microscopic studies was at its peak between 1665 and 1710, but that from the mid-1680s the output in microscopic literature steadily declined. It is also clear that, from 1685 until his death in 1723, Leeuwenhoek alone accounted for about three-quarters of the total production in this specialized branch of scientific activity. Finally, the chart indicates that for microscopy a new period of great activity started in the early 1740s. From the titles in appendices 1–4 it appears that this renewed spurt of activity was to a large extent due to Abraham Trembley's study of the green freshwater polyp and the discovery of its curious properties.

The quantification here of microscopic literature published between 1620 and 1750 thus demonstrates that Hooke, Malpighi, Grew, and Swammerdam began their research and published their observations in the decades during which microscopy was becoming a subject in which many

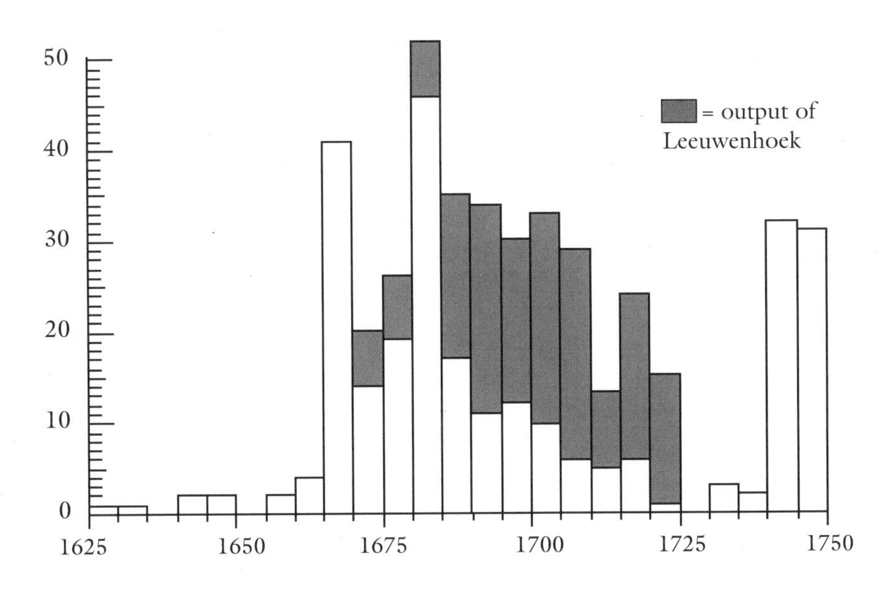

Chart 1. Distribution of publications on microscopy between 1625 and 1750.

people were interested.[91] Leeuwenhoek, however, published mainly at a time in which few people appeared to be concerned with the subject. Hooke and Malpighi, at least as far as Malpighi's earlier contributions were concerned, published their books at a time when hardly any literature on the subject was available. They more or less pioneered microscopy. On the other hand, Grew, Swammerdam, and particularly Leeuwenhoek benefited, or at least might have benefited, from the experiences of their peers as reported in their books.

The Microscope and Microscopy

From the present analysis a few generalizations concerning the relationship between the development and employment of the microscope become clear. The microscope had definitely been invented by 1620. Therefore, by the time microscopy became an important subject, the instrument had already existed for some forty years. In the course of these decades only a small number of microscopes had been produced, and the design of the instrument was hardly developed. Neither the general form nor construction was essentially changed. Suggestions in that direction and actual variations are only seen after the early 1660s. This was the beginning of a period in which numerous additions, alterations, and refinements to the construction of the microscope were initiated, a period resulting in the establishment of the Culpeper tripod microscope as the dominant model during the first half of the eighteenth century.

The train of events was rather similar with respect to the optical parts of the microscope. Prior to 1660 only one notable change in the optical system is noted, namely, the field lens, but after that date the rate of development shifted into a much higher gear. The various suggestions and trials for increasing magnification and diminishing the disturbing effects of spherical and chromatic aberration, however, did not lead to substantial results. The majority of compound microscopes in the first half of the eighteenth century were equipped with a three-lens optical system: eyepiece, field lens, and objective. The specimens of this period were usually furnished with a set of interchangeable objectives, so that the range of magnification in individual microscopes was much extended in comparison with the average instrument of earlier decades. In conjunction with the addition of a number of accessories this meant that the eighteenth-century compound microscope was a much more versatile instrument than its seventeenth-century precursor.

The most important development in microscopic optics was undoubtedly the introduction of the single-lens microscope. The superior optical properties of this type of instrument were immediately obvious, but its development into an easily manipulated tool was not taken up by craftsmen until the 1680s. Leeuwenhoek's discoveries, made with this type of instrument, demonstrably stimulated this development. Consequently the simple microscope was professionally produced in abundance throughout the eighteenth century, but prior to 1680 such an instrument was usually a homemade product.

Judging by the number of publications, the 1670s and 1680s were a period of substantial output in microscopy. In the preceding half century only a few accounts of microscopy had been published.[92] A substantial improvement of the microscope might have offered a plausible explanation for the cause of the sudden rise in microscopy. However, from the above account it appears that such was not the case. Indeed, during the relevant period the optical performance of the compound microscope, expressed in terms of magnification and resolution, did not improve to any great extent. It was not until the introduction of the simple microscope that a substantial improvement in magnification and resolution occurred. As we have seen, the simple microscope did not find wide recognition until the 1680s, by which time microscopic science had long since celebrated its first successes.

From the above survey it is clear that the majority of suggestions for the improvement of the microscope, both optical and mechanical, were launched and executed after microscopy had become well established. The only notable exception was the introduction of the field lens. As this lens, despite some advantages, in fact decreased the performance of the microscope, this development in the optical system cannot have exerted a positive influence over the new use of the instrument. On the contrary, events suggest that the results from microscopic investigation acted as a stimulus for further development of the instrument. Both scientists and instrument makers were involved in improving the microscope. Thus the instrument maker Marshall incorporated the astronomer Hevelius's idea for an improvement of the focusing mechanism in his model, and the instrument makers Butterfield and Wilson adopted designs for simple microscopes from the scientists Huygens and Hartsoeker, respectively.

It appears, therefore, that the development of the microscope did not constitute a substantial factor in the establishment of microscopy. This is supported by the paradoxical fact that microscopy declined precisely at the time when microscopes, both compound and simple, could for the first time be purchased relatively easily from a much expanded body of instrument makers. Whether or not the limitations of the microscope constituted an important factor in the decline of microscopy toward the end of the seventeenth century is a question we shall be able to answer after a detailed examination of the microscopists' exploits.

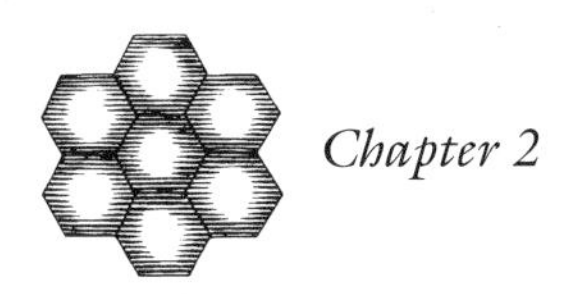

Chapter 2

The Leading Microscopists

IN THIS CHAPTER the objectives underlying the research efforts of the five leading microscopists of the seventeenth century—Robert Hooke, Marcello Malpighi, Jan Swammerdam, Nehemiah Grew, and Antoni van Leeuwenhoek—will be examined in some detail, as well as their perception of the fabric and operation of nature.

Robert Hooke

Robert Hooke's *Micrographia,* which was published on the last day of 1664 although 1665 is printed on the title page, was one of the most famous and important books on microscopy to appear in the seventeenth century. Hooke was one of the foremost scientists of his day, a man whose influence was felt throughout the scientific community of Europe. During his lifetime he published numerous treatises, partly in the *Philosophical Transactions,* partly in separate collections.[1] Even so, a mass of additional material was published after his death.[2] Of the publications he saw through the press himself, *Micrographia* is without doubt his magnum opus, both in volume and scope. This book not only contained a wealth of microscopic facts and figures, it also included excursions into various topics in physics, most of them related to Hooke's theory of light, but also touching on his theory of combustion. *Micrographia* contained a detailed description of Hooke's microscope as well as descriptions of various other instruments such as the barometer and hygroscope.

The Royal Society played a very dominant role in Hooke's life.[3] Most of the salaried positions to which he was appointed, such as Curator of Experiments (1662), Cutlerian Lector in Mechanics (1664), and Gresham Professor in Geometry (1664), were within the Royal Society or closely associated with it. Hooke was also appointed Surveyor of the City of London after the Great Fire of 1666 and devoted a considerable amount of time to activities related to that function, designing and overseeing the construction of buildings, which also brought in a good deal of money.

Hooke's daily life revolved around his rooms in Gresham College, the meeting room of the Royal Society, and the coffeehouses in London. In the latter he met frequently with instrument makers and other craftsmen. Throughout his life Hooke moved to and fro between the world of the craftsman and the world of the scholar. His position within the Royal Society was a somewhat problematical one.[4] He was in fact an employee of the society, to do what the assembled members wanted him to do. Nevertheless he was also a member, even acting for some years as secretary of the society in the mid-1670s.

The Royal Society heavily influenced the course of Hooke's scientific career. In his capacity as Curator of Experiments Hooke was called on to provide topics of interest for the members of the Royal Society at their weekly meetings in the form of experiments and demonstrations, often supplementing these with discourses on a particular subject. He performed these activities with great enthusiasm and ingenuity, thereby upholding the ideal of experimental work of the Royal Society during the 1660s and 1670s.[5] Conceding to the wishes of his fellow members Hooke was forced to turn his attention to an extraordinarily wide range of subjects, including carriages, fountains, pendulum clocks, magnetism, geology, gravitation, music, and respiration. In view of the uninterrupted output that Hooke was to maintain for many years, this function must have been very congenial to his intellect.

Above all Hooke was an inventor and propagandist of scientific instruments. He had a lifelong interest in the advancement of scientific instruments, which in his view were of the utmost importance in science, not only because of the greater number of data they recorded but also because they afforded models by which to interpret nature.

In the preface to *Micrographia* Hooke presented an analysis of the role of microscopes (or rather of scientific instruments in general) in science, in connection with the method of scientific research. Any scientific enquiry, in Hooke's epitome, "is to *begin* with the Hands and Eyes, and to *proceed* on through the Memory, to be *continued* by the Reason; nor is it to stop there, but to *come about* to the Hands and Eyes again."[6]

This is of course the very essence of the Baconian program, to which Hooke was wholeheartedly committed. In the context of this program

Hooke intended his *Micrographia* as a demonstration that the new science above all needed "a sincere Hand, and a faithful Eye, to examine, and to record, the things themselves as they appear."[7] The senses, as they provided "intelligible, rationall and true"[8] explications of the phenomena of nature, stood at the center of Hooke's concerns in science. To Hooke, the quintessence of a scientific instrument was making good the inadequacies of the human senses; an instrument could thus be designated as an "artificial Organ." The extent to which the normal range of eyesight could be enlarged with both the telescope and microscope had been abundantly demonstrated in what were then recent times. Hooke did not doubt that the senses of sound, smell, taste, and touch could equally well be supplemented with appropriate aids.

The term "artificial organ" might, however, more appropriately be applied to those instruments registering the operations of nature that lie outside the scope of human senses. Such instruments, wrote Hooke, "may be made a Genus, as it were, of new Sorts of Sense, comprised under them, of which we have yet no Notion, nor any Sense or Method of Discovery; at least they are yet unheeded."[9]

Hooke's ingenuity was particularly suited to the task of devising such instruments, among them the barometer and hygroscope described in *Micrographia*. One of the reasons for publishing *Micrographia*, wrote Hooke in the preface, was to promote the use of instruments in science. Where the microscope was concerned, he was convinced that "by these helps the subtlety of the composition of Bodies, the structure of their parts, the various texture of their matter, the instruments and manner of their inward motions, and all the other possible appearances of things, may come to be more fully discovered."[10]

In the course of his career Hooke often worked on the development of new instruments or the improvement of existing ones. His ideal was

> to improve and increase the distinguishing faculties of the senses, not only in order to reduce these things, which are already sensible to our organs unassisted, to number, weight, and measure, but also in order to the inlarging the limits of their power, so as to be able to do the same things in regions of matter hitherto inaccessible, impenetrable, and imperceptible by the senses unassisted. Because this, as it inlarges the empire of the senses, so it besieges and straitens the recesses of nature: and the use of these, well plied, though but by the hands of the common soldier, will in a short time force nature to yield even the most inaccessible fortress.[11]

To Hooke there was more to instruments than just the examination and measurement of natural phenomena:[12] the mechanics underlying their construction also provided models with which to interpret the same phenomena. The mechanical philosophy's doctrine holds that the subvisible world is made up of structures comparable to those in the macroscopic world and that similar forces work on them to bring about the processes observed with the naked eye. Accordingly Hooke insisted that for any effect whose cause is invisible it is necessary to make an analogy with a similar effect whose cause is visible. On several occasions he therefore built models that were to demonstrate and thereby prove mechanisms for subvisible structures: a case in point is the contraction of muscles, which will be discussed in chapter 4.

At about the same time Hooke composed *Micrographia,* he also began a treatise entitled *A general scheme . . . of the present state of natural philosophy,* which was not published until 1705, when Richard Waller included it in *The posthumous works of Robert Hooke.* The uncompleted manuscript contained Hooke's views on the first step in scientific enquiry, that of the empirical gathering of data. It had been his intention to complete the manuscript with a discussion of the second step, that of reasoning. He claimed to have developed a method, which he referred to as his "Philosophical Algebra," to serve as an instrument in that process. Historical analysis has revealed Hooke's Philosophical Algebra to be a method for successively raising hypotheses, which may be tested against the phenomena.[13] Hooke claimed that "it is possible to do as much by this method in Mechanicks, as by Algebra can be perform'd in Geometry."[14] In *Micrographia* many traces can be found of Hooke's method, the most conspicuous being the sequence of the various topics. "As in Geometry," wrote Hooke, "the most natural way of beginning is from a Mathematical *point;* so is the same method in Observations and *Natural history* the most genuine, simple and instructive."[15] Accordingly the first observation is that of a point (of a needle), the second of a line (the edge of a razor), which is followed by several planes (pieces of woven material) and spatial objects (particles of sand and minerals), and finally by a series of vegetable and animal objects.

However, the arrangement of his observations according to geometric principles was, at this level, intended for instructive purposes rather than the outcome of the prosecution of a specific enquiry. Yet in a number of "Observations" Hooke's method is apparent in the models he proposed for further enquiry into a specific detail. A case in point is the observation concerning "blue mould."[16] Hooke presumed the growth of molds to be a mechanical process, that is, the heat of the air excites appropriate materials of putrefying bodies into vegetation. The description of the observed specimen, according to the author, should have been followed by a survey of the multitudinous forms found among different kinds of mold and an enquiry into their manner of growth and vegetation. The latter enquiry should have proceeded from the following considerations: (1) mold requires no seminal property to propagate, (2) molds produce no seeds, (3) bodies may be generated from saturated solutions, actuated by heat, (4) mushroom-shaped excrecencies are formed at the top of the snuff of candles, and (5) on the roof of vaults mushroom-like "icicles" are formed out of "lapidescent" water. On the basis of these considerations Hooke posited that mushrooms are formed as an impregnated liquor ascends, evaporates, and leaves behind its more solid parts. Before definitely deciding on such a mechanism Hooke felt that there should be an inquiry into the materials of which this liquor was composed and what caused it to ascend, and whether other principles acted as agents or aids in the process. Even though Hooke did not himself execute these steps of his Philosophical Algebra in this case, he was convinced that it would prove that the growth of these simple forms of vegetable organisms came about in a manner akin to the growth of salt crystals. Indeed, he could not "find the least possible argument to persuade me there is any other concurrent cause than such as is purely Mechanical."[17]

Besides presenting a survey of microscopic observations on a wide range of subjects, Hooke also used his *Micrographia* as a vehicle for a discussion of his views on the organization of matter into mineral, vegetable, and animal substances. For the purpose of that discussion Hooke investigated and discoursed on simple and uncompounded bodies before turning to the more complex ones. Hooke's selection of objects and the order of their insertion in *Micrographia* was thus to a certain extent affected by his intention to discuss his conception of nature as a corollary to his observations. Indeed, "these several Enquiries having no less dependance one upon another than any select number of Propositions in Mathematical Elements may be made to have."[18]

The sequence of observations was determined by the way in which Hooke's argument was constructed on the thesis that the increasing complexity of the organization of matter was caused by (successively) the principles of "fluidity, orbiculation, fixation, cristallization, ebullition or germination, vegetation, plantanimation, sensation and imagination."[19] These terms described the successive processes that in Hooke's view fashioned substances of more complex structure from the simpler units of the preceding level of organization. Accordingly,

> Air and Water [have] no form at all, unless a potentiality to be formed into Globules; and the clods and parcels of Earth are all irregular, whereas in Minerals (nature) does begin to *Geometrize,* and practise, as 'twere, the first principles of *Mechanicks,* shaping them of plain regular figures, as triangles, squares, &c and *tetraedons,* cubes &c. But none of their forms are comparable to the more compounded ones of Vegetables; For here she goes a step further, forming them both of more complicated shapes, and adding also multitudes of curious Mechanick contrivances in their structure.[20]

In *Micrographia* the "property of congruity," which causes the particles of unorganized matter to cohere, and the crystallization of globules into regular forms are discussed at length.[21] These principles concerned nonliving matter. As far as living matter was concerned Hooke discussed the mechanical growth or germination of simple vegetable organisms, as discussed above, but the subsequent principles of organization in living matter were, except for a few hints, left out of consideration. The examination of a hair from the beard of an ear of oats prompted Hooke to make a few allusions to animate matter. The movement of this vegetable structure seemed to Hooke "to be the very first footstep of *Sensation,* and Animate motion, the most plain, simple, and obvious contrivance that Nature has made use of to produce a motion."[22] Apropos of this conjecture Hooke speculated that, just as the hair of the oat's beard causes movement because it absorbs moisture, so may the contraction and relaxation of muscles be caused by the influx and subsequent evaporation of some kind of juice.[23]

Hooke did not carry his speculations concerning living matter beyond this level of organization in *Micrographia.* However, some twenty years later in a posthumously published collection of lectures read to the Royal Society concerning the nature, properties, and effects of light,[24] Hooke set down his views concerning the action of the brain. Central to his explanation was the premise that the brain contains five distinct substances, each adapted to receive and store the impressions of one of the five senses. Therefore, human mental processes were conducted with memories and ideas that were "material and bulky, that is, to be certain bodies of determinate bigness, and impregnated with determinate Motions."[25]

Nowhere is Hooke's mechanical view of nature more pronounced than in his discussion of the generation of primitive plants and animals.[26]

He thought it very possible that these simple organisms are produced by a chance reassembling of the broken-down parts of more complexly structured plants or animals—that is, an accidental production. Hooke imagined that as plants and animals decay, the constituent parts of their bodies become dislodged, but some of these parts remain whole and can be recycled into a new organism. Hooke elucidated his views by drawing a parallel with the mechanism of a clock: by removing some parts of a complicated but broken-down clock, the remainder may run again, although not on time. A further implication of his view that animate matter is expressed in a variety of "mechanical contrivances" was, in Hooke's opinion, that different organisms may be composed from identical parts and, conversely, that like organisms may be constructed from different parts. Apropos of the multitudinous forms of insects Hooke surmised that "the All-wise God of Nature, may have so ordered and disposed the little *Automatons,* that when nourished, acted, or enlivened by this cause, they produce one kind of effect, or animate shape, when by another they act quite another way, and another Animals is produc'd. So may he so order several materials, as to make them, by several kinds of methods, produce similar *Automatons.*"[27]

Marcello Malpighi

Marcello Malpighi devoted a lifetime of research to the elucidation of life's processes by means of dissection, observation, and experiment.[28] In this enquiry the microscope was one of the principal tools. The main content of Malpighi's scientific research matched the general emphasis placed by anatomical investigation in the seventeenth century on the texture of the organs. Several outstanding anatomists such as Francis Glisson, Thomas Willis, Niels Stensen, and Joseph-Guichard Duverney, who published, respectively, anatomical treatises on the liver, brain, muscles, and the human ear,[29] painstakingly investigated the finer structures of these organs as the sound structural basis for the explanation of their operations. Malpighi's work stands out among his contemporaries because, while like them he wanted to discover the seat of the physiological processes, he was aiming particularly at the resolution of the microscopic details in the texture of organic matter. He thus applied the anatomical method in physiology on a level far below that for which the aforementioned anatomists were aiming.

The direction of Malpighi's investigations, as well as their theoretical framework, was set during his three-year residence (1656–59) at Pisa University as a professor of theoretical medicine. In Pisa, Malpighi formed a close relationship with Giovanni Alfonso Borelli, one of the leading members of the Accademia del Cimento during its peak period from 1657 to 1667. In an autobiographical essay published in his *Opera posthuma*, Malpighi relates that, in the course of his frequent encounters with Borelli, the latter impressed on him the value of the experimental method in scientific enquiry, as well as that of the mechanical conception of nature.[30] While taking part in Borelli's anatomical research, which was to serve as an experimental basis for his *De motu animalium* (1680–81), Malpighi was gradually instilled with Borelli's views on the vital operations of living bodies. Essentially, Borelli held that the phenomena of life are brought into being by the mechanical actions of the subvisible machinery that composes the texture of living bodies.[31] Consequently, the investigation of the physiological processes of human beings and other organisms amounted to the discovery and explanation of that machinery.

After his sojourn in Pisa, Malpighi settled in Bologna in 1659, where he spent the next thirty years of his life, apart from a four-year interlude between 1662 and 1666 at Messina. He was engaged by the University of Bologna to occupy the chair in practical medicine. Three years before his death he moved to Rome to serve as chief physician to Pope Innocentius XII.

The scientific climate in Bologna at that time was a mixture of traditional doctrines and the new, experimental approach to science.[32] Officially, the professors of the university taught their students the conventional curriculum. However, a number of Bolognese scientists were attracted to the Baconian approach to science, as expressed in the activities of the Royal Society in London, and endeavored to practice this novel approach outside the university in scientific societies. Malpighi shared their enthusiasm and, like some of his colleagues and friends in Bologna, was a corresponding member of the Royal Society and sought to contribute to its aims by keeping that learned body informed of recent developments in science in Bologna, and Italy generally.

The Royal Society in its turn has been of vital importance for the dissemination, and consequently the impact, of Malpighi's scientific oeuvre. Having elected him as a member in 1668 the members, through Henry Oldenburg, stimulated him to follow up and record his observations on the anatomy and development of the silkworm, and his research concerning plants. When they received the fruits of Malpighi's labors in the form of manuscripts containing his treatises on the silk moth, the chick embryo, and the anatomy of plants, the members also endorsed their publication. In later years the Royal Society also published Malpighi's *Opera omnia* (1686) and arranged for the publication of various unpublished papers that Malpighi had entrusted to it, papers that in effect constituted his intellectual autobiography, in the *Opera posthuma* (1697). The recognition of Malpighi's achievements by the Royal Society was consolidated by the inclusion of several of his treatises in Le Clerc and Manget's *Bibliotheca anatomica* (1685), which purported to be a collection of recent seminal contributions to anatomy.

Despite his international prestige, Malpighi was harassed in Bologna by the publication of several offensive discourses by his colleagues from the medical faculty.[33] Several of these men supported strictly traditional points of view and attacked Malpighi because of his "modernism," especially because he employed the experimental method in medicine. Giovanni Girolamo Sbaraglia, for instance, in his *De recentiorum medicorum studio dissertatio* (1689) fulminated against the uselessness of comparative anatomy, microscopic anatomy, and the study of plants with respect to the advancement of medicine. Instead, he wrote, attention should be devoted to the careful observation of the sick.[34] Similar viewpoints were voiced by physicians of international repute, such as Thomas Sydenham,[35] but Malpighi was chagrined by Sbaraglia's attack because the latter urged his points exclusively through a systematic demolition of Malpighi's scientific achievements.

The object of Malpighi's scientific research was the elucidation of the physiological processes of living bodies, particularly of the human body. The strategy he employed to achieve his objective consisted of observation, experiment, and reason. Malpighi's enquiries concerning animal physiological processes invariably entailed a detailed study of the structure of the organ involved in a range of animals such as fish, frog, tortoise, fowl, and various domestic animals. "We borrow illumination," he wrote, "as if by degrees, from dissection, sometimes of insects, sometimes of perfect animals. For Nature is accustomed to rehearse with certain large, perhaps baser, and all classes of wild [animals], and to place in the imperfect the rudiments of the perfect animals."[36]

It was therefore a perfectly reasonable procedure, one that was moreover deeply rooted in tradition, to clarify human anatomical structure with comparative studies of animal anatomy. In his physiological treatises Malpighi achieved this—his first—approach toward the resolution of minute structures, mostly by means of the dissection of various vertebrates, observing as it were the texture of human organs through the microscope of nature.[37] In subsequent studies and publications Malpighi, who realized that until then his animal studies had not penetrated into the ultimate structural schemes of living matter, actually incorporated the anatomy of the insect body and especially the structure of plants in his analogical method. In fact, he wrote, "precedence is due to even a more simple world, that of the minerals and elements,"[38] but even Malpighi's mind boggled at the enormity of that task.

Malpighi's second approach to the texture of living matter comprised the application to the objects of various techniques of anatomical preparation, comprehended under the term *anatomia subtilis.* Such techniques included the injection of colored liquids into the vessels, insufflation of the vessels, desiccation, cooking, and maceration of the excised materials. A novel feature of Malpighi's approach was the combination of anatomical techniques with microscopic inspection. This particular combination allowed the resolution of many details in the texture of animal organs, such as the Malpighian bodies (glomeruli) of the kidney, the capillaries of the blood system in the frog lung, and the structure of the Malpighian (reticular) layer in the skin. The description of these, the more minute, parts of the living body was in Malpighi's view the true object of anatomy, because they formed the basis of all physiological processes.

Physiological experiments were a third approach practiced by Malpighi. His experimental data in animal physiology were to a large extent a result of the anatomical preparation of the objects rather than truly physiological experiments. Noting the distribution of colored fluid in the kidney, as it was injected through the artery, vein, or ureter, Malpighi drew conclusions about the connections between the blood vessels, the spherical bodies (glomeruli) which he regarded as glands, and the renal tubuli. Only infrequently did he complement his anatomical observations with physiological experiments involving vivisection. For instance, he endeavored to demonstrate conclusively the connections between the various structures of the kidney by ligaturing the ureter and renal vein in a dog and sometime later dissecting the excised kidney, without positive results, however.

Malpighi employed these three approaches to arrive at "the exact and truthful description of the fluid and solid parts composing the body, with their position, shape, connection, movement and use,"[39] which required the exploitation of the full potential of the means available. Despite the scores of novel details he discovered with the microscope Malpighi, who repeatedly and convincingly argued in favor of the use of the microscope in anatomical research, often felt frustrated by the limitations imposed by that instrument. Sometimes he found that "structures escape the senses on account of their exquisite smallness,"[40] even when viewed through the microscope. Therefore, Malpighi thought that he might have achieved more if he had only had better optical instruments at his disposal, as he wrote by way of excuse in the preface to his *Anatomes plantarum pars prima* (1675). However, in other cases he found that the qualities of the objects, their transparency, whiteness, and mucosity, interfered with perception. Writing about the embryo of the chicken, for instance, he said, "Everything is so mucous, white and transparent that the eye, with whatever instrument it may be fortified, is unable clearly to detect the contexture of the parts."[41] It was therefore not so much the rather small magnification (by the standards of today) of contemporary microscopes that determined the limitations of Malpighi's microscopic research, but rather the lack of adequate microscopic technique to get around the obscuring details of the object.

Just as Malpighi relied on comparative anatomy to exemplify details invisible in human objects, he resorted to analogy in cases where he was struck by the inadequacies of microscopic inspection or by the limits of anatomical artifices. Details that eluded the "eye of sense," he felt, could well be established with the "eye of reason." Thus several crucial inferences in Malpighi's expositions were the result of analogical reasoning. Although he could not, for instance, demonstrate a continuation between the renal tubuli and the "glands" (glomeruli), he concluded that such a continuation must exist, for "if in the liver, in the brain, and in other glands, it is the invariable rule that each single acinus, or globule of the gland throws out its own excretory duct besides the arteries and veins, the same will have to be said about glandular bodies of this sort."[42]

Malpighi therefore combined empiricism with analogical deduction on two different levels. One concerned the visible and subvisible pattern of the texture of specific parts such as the lung and liver and was founded on comparative anatomy. The second, more fundamental level, concerned a uniform pattern of organization that originated from his microscopic observations and was supported by his conception of nature.

At several points in the above discussion of Malpighi's microscopic investigations, the fundamental assumption on which his scientific work was founded has become apparent, namely, the principle of uniformity in nature—uniformity both in the structure of organic matter and in the mechanism of its operation. Allowing that all animate matter was constructed from identical materials, or *minima* in Malpighi's vocabulary, the scheme of construction in animals and plants was found to have few features in common. The structural similarity between the trachea in insect bodies and the spiral vessels in plants was an exception rather than the rule, and Malpighi was therefore delighted with this discovery. Structural resemblance signified similar function in his view, and therefore the spiral vessels must also serve to transport air. Moreover, the air within the animal body and the plant served similar functions in that it supported nutrition in both cases, in animals by mixing the newly absorbed nutriments with the blood, and in plants as a driving force for the mechanism of sap transportation.

Malpighi used the principle of uniformity of construction to the best advantage in his comparative studies of human and animal organs. Thus he decided by argument from analogy with the lungs of frogs, that numerous minute blood vessels run through the walls of the vesicles of human lungs, even though he could not visually establish their presence. In his view uniformity of the fundamental plan of construction was not limited to a particular organ throughout animal nature, but also pertained to the various organs. In the several discourses collected in *De viscerum structura* Malpighi referred frequently to similar plans of construction he had observed in other organs. The partitioned structure of the spleen, for instance, seemed to him to have much in common with the structure of the lung as observed in the tortoise. Similarly, he likened the structure of the brain's cortex with that of the liver.[43] The proposition of a uniform plan of construction corresponds with Malpighi's view that the principal actions performed by the various organs are the separation and mixture of particles.[44] Both processes are brought about by the *minima,* or smallest organized units, the former by forcing the particles through a capillary tube and the latter by means of a sievelike action. Since Malpighi understood the function of the kidney, liver, and the cortex of the brain to be the separation of urine, bile, and "nervous juice," respectively, he envisaged the structure of these organs to be particular adaptations from a uniform scheme, that of the glandular follicle.

Mixture and separation, therefore, were brought about by mechanical means. Malpighi was a confirmed mechanist and wrote, "The machines of our body . . . are made up of cords, filaments, beams, levers, tissues, fluids coursing here and there, cisterns, canals, filters, sieves, and similar mechanisms. Man, examining these parts by means of dissection, philosophy, and mechanics, has learned their structure and use; and, proceeding a priori, he has succeeded in forming models of them, by means of which he demonstrates the causes of these effects and gives the reason for them a priori."[45]

Even though Malpighi enumerated a number of such models, among them an optic camera as an analogue to the eye, the levers and sieves in the above quotation must be understood as a metaphor of the mechanical origins of physiological processes rather than as an accurate description of the manner of their operation. Malpighi phrased this passage in words very similar to those used by Borelli in the preface to his *De motu animalium,* who mentioned as examples for the mechanical causes of the operations of animals the scale, lever, pulley, tympanum, wedge, and the screw.[46] Apparently a certain measure of homage to Borelli has affected Malpighi's choice of words in which to phrase his mechanical conception of nature, the more so as Malpighi was very well aware of the fact that he had not penetrated the texture of animate matter into its finest details, and that further details lay tantalizingly just beyond the limits of the resolution of his microscope. He wrote to one of his correspondents, Johann Theodor Schenk (the author of a treatise entitled *De poris corporis humanis,* published in 1670), that he found it impossible to discern with the microscopes the smallest "pores and passages" in the tissues. He felt sure that the search for such details would be vain and that consequently the greatest secrets of nature would remain hidden.[47] Consequently Malpighi expressed his views on the operation of physiological processes more cautiously in relevant discourses than in the statement above concerning his allegiance to the mechanical philosophy. With respect to the action of the glands in the kidney, for instance, he deliberated that

> [it is] reasonable to assume that this [the production of urine] is wholly the result of the work of the glands but since the minute and simple structure of the openings within the glands escape us, we can only postulate some things in order to give a satisfactorily probable answer to this question. It is obvious that this mechanism accomplishes the work of separation of the urine by its internal arrangement. But whether this internal arrangement is similar to those devices which we make use of here and there for human needs, and in imitation of which we build rough contrivances, is doubtful. For although similar sponge-like bodies, structures with sieve-like fistulae, may be encountered, it is difficult to determine to which of these the structure of the kidneys is similar in all respects. And since the manifestation of nature's working is most varied, we may discover mechanisms which are unknown to us and whose operations we cannot understand.[48]

Malpighi, therefore, positively asserted mechanical forces as the modus operandi in physiological processes. However, he estimated that his observations, even though they abundantly confirmed the complexity of the structure of organic matter, could not be reconciled to rather crude mechanical models. Therefore if the physiological processes were brought about by mechanical actions these did not take place on the level of structure resolved with low magnification.

Jan Swammerdam

Swammerdam's personal life was anything but harmonious.[49] His passionate devotion to scientific investigation was at the root of the most severe and prolonged of the various conflicts that disrupted his life. One of these conflicts arose when Swammerdam preferred to study medicine rather than theology, a choice to which his father apparently only grudgingly consented. This conflict spoiled his relations with his family, particularly with his father, from his student days until his father's death. As a student at Leiden University Swammerdam did not restrict himself to the prescribed curriculum but started to experiment on his own. Some of his experiments are described in the diary of Olaus Borch, a professor from the University of Copenhagen, who stayed for a few months in Leiden during a prolonged tour of various European universities.[50] Swammerdam's future interests and skills are already apparent from Borch's entries. On various occasions he displayed his cabinet of insects to Borch, performed some experiments with live dogs, and demonstrated, in connection with Malpighi's recently published *De pulmonibus,* the alveoli of a frog's lung.

Despite these promising indications, Swammerdam delayed the formal completion of his studies by absenting himself for three years from Leiden University. Part of this period was occupied by a prolonged sojourn in France. When he had finally become a doctor of medicine in 1667 by presenting and defending a doctoral thesis entitled *De respiratione,* Swammerdam settled in Amsterdam. There, he did not take up the practice of medicine in order to earn a living, but devoted his time to scientific interests, living at his father's expense. This state of affairs caused such tension between father and son that Swammerdam once wrote to his friend Melchisedec Thévenot that he was going to give up "anatomy, insects and all curious experiments," being forced to that decision because his father was "no more inclined to provide . . . money or clothes."[51] However, his father apparently relented, as Swammerdam continued to live in his father's house and went on with his research without interruption. In order to ensure financial independence from his father, Swammerdam eventually decided to sell his cabinet of natural curiosities. In 1668 no less than 12,000 guilders had been offered for that collection (after which time the collection was again substantially enlarged),[52] and as Swammerdam estimated that he only needed 400 guilders a year to sustain him, the sale of his cabinet would definitely resolve his difficulties. However, the sale of the cabinet did not come off nor did the sale of his father's cabinet some years later. The Swammerdam siblings quarreled over their father's estate, so that the sale of Swammerdam père's cabinet, which would have meant a future free from financial worries for his son, did not take place until after Swammerdam had died. In sum, Swammerdam's relations with his family were strained throughout his adult life mainly because he wanted to devote his time to scientific investigations rather than earn his own living.

A second conflict raged in Swammerdam's own mind between his passion for science on the one hand and his deep religious feelings on the other. Nearly every other page in his *Biblia naturae* testifies to his belief that God's omnipotence is nowhere more visible than in the intricate structure of minute living beings. A typical declaration of his feelings is the following paragraph:

> Look, so all-wonderful is GOD, in respect of these small animals, so that I dare say, that in the insects GOD'S countless wonders are sealed up, which seals are revealed as one diligently turns over the leaves of the book of Nature, the Bible of Natural Theology, in which GOD'S invisibility becomes visible; because treasuries of ineffable wonders then manifest themselves; and the hidden Creator becomes so manifest in these small Animals, that the experiences of the same, serve me as the biggest proofs to evince without yielding his eternal Divinity and Providence against all his detractors.[53]

Swammerdam conceived the study of nature as an exploration and confirmation of God's glory, and thus as a kind of divine worship. However, his investigations were also a time-consuming occupation, which prevented the giving of due attention to traditional forms of worship. Over the years Swammerdam came to feel that, by indulging in scientific research, he was neglecting his vital duties as a Christian. In the preface to *De ephemeri vita,* dated 12 July 1675, he wrote, "I have now spent enough time and labor in the investigation of Nature and have followed my own depraved will and pleasure therein. Wherefore I now intend to follow solely God's will, to surrender my will to Him, and withdraw all my thoughts from the multiple things so as to offer them to heavenly reflections only."[54] Swammerdam was obviously trapped in a crisis, struggling with conflicting desires. In the end this resulted in a decision to renounce scientific research and to join the religious community of Antoinette Bourignon. This decision was supported by his feeling that his investigations until that time had already guided him to God and the continuation of that research would only impede his worship of God.[55]

However, from the circumstances that delayed his departure to Bourignon's side it is clear that Swammerdam's mind was still inclined to science. He first saw his treatise on the mayfly, entitled *De ephemeri vita,* through the press. This book is a perfect reflection of his state of mind at the time. It is a mixture of a superb study of the life and anatomy of the mayfly, of lamentations on the futility of human life, of prayers, and of digressions into theological questions. Before he left Amsterdam Swammerdam also went through the notes of his researches and destroyed some of these, among them his notes on the anatomy of the silkworm.[56] However, most of his notes were still extant when he resumed his scientific activities about a year later, and he took care that his drawings of the silkworm's interior parts were sent to Malpighi, so that the results of his work would not be lost.

During his stay in Bourignon's community the conflict between science and religion was to some extent resolved in Swammerdam's mind. On his return to Amsterdam, he devoted all his time to editing his notes, which he elaborated and completed and which were supplemented with a series of newly initiated investigations.[57] Eventually all of this was to be published as the *Biblia naturae* through the good offices of Herman Boerhaave, half a century after the author's death. Swammerdam therefore executed this research, as he frequently stated in personal remarks inserted between detailed descriptions "solely to the Glory of GOD and without any other intention."[58] Even though this was a deeply felt sentiment, the connection between his religious and scientific passions was certainly not only stimulating and fertile,[59] but also competitive and, during at least one period in his life, destructive.

The trying circumstances of Swammerdam's private life were balanced by a number of supportive friendships, including those with Niels Stensen and Melchisedec Thévenot. Stensen's life was in several respects a parallel to that of his friend, as he was similarly torn between science and religion.[60]

Unlike Swammerdam, however, Stensen resolved to follow a professional career in religion. Swammerdam and Stensen first met in Leiden when both were students, and they collaborated in the investigation of muscles.[61] In subsequent years they met again in Paris when both were drawn into the circle of scholars surrounding Thévenot, which was to become a foundation stone of the Académie des Sciences. Although Swammerdam met both Stensen and Thévenot again in later years, these encounters were only brief, so that most of the scholars' exchanges were enacted through letters.

In Leiden and in Amsterdam Swammerdam found opportunities to perform his research in congenial company. In Leiden he studied with Stensen and Reinier de Graaf under Sylvius, and later worked with Johannes van Horne. In Amsterdam he joined the *Collegium privatum Amstelodamense*, and he dominated the activities of this small body of men, Gerard Blaes also prominent among them.[62] Although Swammerdam was not perturbed when some of his most accomplished anatomical feats, such as the demonstration of the valves in the lymphatic vessels and the structure of the spinal marrow, were published under other men's names (respectively, Frederik Ruysch[63] and Gerard Blaes[64]), he reacted rather uncharacteristically with respect to de Graaf's *De mulierum organis* (1672). In his polemical *Miraculum naturae* (1672), which he published directly on the publication of de Graaf's book, Swammerdam claimed priority for van Horne, Stensen, and himself concerning the discovery of certain details of the human ovaries, which he had already published using an engraved plate the year before.[65] This untypical action on the part of Swammerdam may have been prompted by feelings of friendship for Stensen and gratitude for van Horne's patronage. Be that as it may, his claim was rejected by the Royal Society with whom he had lodged it.[66]

Swammerdam was wholeheartedly committed to the empirical method in science. He regarded observation and experience as the prerequisites for any certain knowledge of nature. Knowledge derived solely from books was worthless in his view, and he fulminated against deductive reasoning. Many a philosopher, he wrote, had erred to a distressing degree by relying on reasoning while forgetting to observe the phenomena in the first place,[67] an assessment that was substantiated with a reference to the faulty notions of metamorphosis he had encountered in the relevant literature. Swammerdam's published research demonstrates that he kept strictly to his own rules and rarely advocated views that were not supported by the results of his own observations and experiments.[68] He observed, experimented, described, and reached conclusions that had, by way of a final test, to be checked against nature. Indeed,

> When our reason is false and wanting; when she cannot be supported by experience, cannot be proved by it, and does not terminate in the same, then it seems to me, that there can be no stronger or more powerful reasons than those which are extracted from experience and practice, in which they must end. All other reasons, which do not have this firm and unmovable foundation, no matter on how many inductions and conclusions they rest, must be regarded with some suspicion, and if they do not accord with experience, they must be discarded entirely.[69]

To Swammerdam the outcome of the process of amassing data and of subsequent induction was a clearer insight into the laws and order prevalent in nature. Although he conceded that, in principle, man might discover causal explanations of the phenomena, in practice he deemed this to be impossible, partly because of the feebleness of man's mind but also as a result of the limitations of man's powers of observation. He argued that

> just as we cannot obtain true experience of all things, and have therefore no clear and distinct notion of the same (like those which are too small for our vision and like others that are too far removed from it), just so we should not foolishly imagine that we shall ever obtain through our reason true and real knowledge of the causes of things, let alone of her true manifestations. Our greatest wisdom lies . . . not in knowledge concerning the causes of things but only in a clear and distinct notion of [nature's] true manifestations or effects.[70]

Swammerdam found the structure of matter inscrutable; indeed, "eye, hand, reason, and instruments together are, because of its minuteness, too impotent"[71] to discover the finer details of its structure. Thus man's knowledge must of necessity remain limited: the ultimate causes of the phenomena stay hidden and can only be known to God. Therefore Swammerdam judged the accurate and detailed description of nature as the highest goal of scientific enquiry.

These views on scientific method and knowledge appeared in the last chapter or epilogue of Swammerdam's *Bloedeloose dierkens* of 1669. It is interesting to note that, in this chapter, Swammerdam implicitly dissociated himself from views he had advanced some years before in his doctoral thesis, entitled *De respiratione*, which had been published in 1667. In his thesis Swammerdam presented a series of brilliantly executed experiments and vivisections, with the object of proving that the air in respiration is not *attracted* into the lung as a result of a partial vacuum, but is rather *pushed* into it as a result of the expansion of the chest. This point was of great importance since Swammerdam's objective was to reconcile the mechanism of respiration with the notion that all movement is caused by collision between particles and has nothing to do with attracting powers.[72] These particular investigations were therefore designed to provide experimental evidence for a mechanical explanation of respiration, which was indeed thoroughly discussed by Swammerdam. As only two years later he rejected the possibility that man may arrive at sure knowledge about the causes of such processes, this attempt at causal explanation in *De respiratione* appears to constitute an anomaly within the total of his scientific output. This may be attributable to the influence of Swammerdam's associates in Leiden, and particularly to the scientific circle surrounding Franciscus dela Boë Sylvius, who was Swammerdam's intellectual mentor at Leiden University.

A second series of physiological experiments, performed by Swammerdam at about the same time, appears to support the view that his mechanical explanation for respiration was inspired by his peers at Leiden University rather than being based on his own judgment. Using a nerve-muscle preparation of a frog's hind leg Swammerdam demonstrated that the muscle, contrary to current belief, does not increase in volume during contraction.[73] These experiments were not published at that time, although they were known among his acquaintances as will be seen in chapter 4, but only appeared in print half a century after his death in his *Biblia naturae*. In his account of the experiments Swammerdam effectively destroyed the theory that the muscle contracts as the result of an influx of matter from the brain, but he did not produce an alternative explanation, presumably because of his deeply felt conviction that God's works are ultimately inscrutable. Even so, from his researches on respiration and muscle contraction, it is apparent that during his formative years in the 1660s Swammerdam was involved with the contemporary shift in physiology from traditional explanatory notions toward a mechanical explanation of the various processes.

By the time his *Bloedeloose dierkens* was published Swammerdam had come to despair of ever arriving at causal explanations and exclaimed, "O GOD, Thy Works are inexplorable, and all we know, or can know of them, are nothing but the dead shadows of the shadows of shadows of Thy adorable and inexplorable works; for which all the minds of man, however ingenious they may be, must become dull and confess their ignorance."[74]

From then on he was content to point out "the rules and order, which the all-wise Creator has instilled unchangeably in the nature of things."[75] He considered that he had contributed substantially to science by discerning four different types of metamorphosis among the mass of observations on the development of numerous individual insects.[76] Nevertheless, on occasion Swammerdam suggested a mechanical explanation for the phenomena he observed. For example, he attributed the hardening of the almost liquid parts of the butterfly within the body of the caterpillar to the evaporation of water.[77] Most of his suggestions, however, were equally insubstantial and therefore hardly suffice to represent him as a mechanicist.[78] With regard to Swammerdam's microscopic work this means that he made no attempt to solve any problems concerning the operation of living beings, as Hooke and Malpighi had tried to do. Having turned his mind and experimental skill toward the study of insects, Swammerdam set out to destroy the current notions concerning this group of animals, which he found to be completely mistaken.

These notions are, first, that insects propagate through spontaneous generation; second, that in the course of their development they change suddenly from one form into another; and third, that they lack any internal structure. The notion of spontaneous generation was categorically dismissed by Swammerdam in view of his belief in the uniformity of nature, which precluded chance,[79] and chance is exactly what may occur in spontaneous generation. A careful study of the development of a variety of insects, published in 1669 as *Historia insectorum generalis, ofte, algemeene verhandeling van de bloedeloose dierkens,* revealed the various types of metamorphosis and established beyond a shadow of doubt that this process was one of gradual change.

Having settled this matter, Swammerdam turned toward the investigation of the anatomy of insects. It was an absorbing study, executed with great skill, but most of it was not published until half a century after his death, when Boerhaave published Swammerdam's manuscripts as *Biblia naturae* in 1737 and 1738. It is this part of Swammerdam's researches that was to a large extent performed with the microscope. The results of his work were initially incorporated into separate treatises and letters to friends, such as Thévenot, although it was Swammerdam's intention to collect these into one volume.[80] The tracts and letters are not of an argumentative kind. On the contrary, they are purely descriptive, setting out in painstaking detail the life cycle and behavior of the various animals, and the arrangement and function of their external and internal parts.

Swammerdam's conception of nature was based on rigorous order, a concept that precluded chance and corresponded with uniformity.[81] Consequently, he rejected both the notion of spontaneous generation and the contemporary view of the metamorphosis of insects as a process of sudden change from one image into another. His exposure of the future butterfly, which lies hidden in a completely finished form beneath the skin of the caterpillar, was a most convincing demonstration with respect to his theory of metamorphosis. Even better, perhaps, this demonstration may in the first place have been decisive in the formation of his ideas on this topic. It is certain that Swammerdam performed this feat in 1668, but probably even earlier, in 1662.[82] In essence, he stated that although the insect changed successively from one form into another it remained the same individual throughout the process. The various parts, which are present in the imago but not in the caterpillar, do not appear suddenly but "grow on slowly, one part after the other . . . and they are increased and born in this swelling, budding forth, rising up, budding and as if stretching of new limbs, gradually by an addition of parts."[83]

Swammerdam's insistence that the insect had already acquired its form in previous stages and that the egg is "the animal itself," coupled to a reference concerning the preexistence of man as far back as the ovaries of Eve and some similar remarks,[84] has caused a number of historians of science to call him a preformationist.[85] As Swammerdam had managed to see some, albeit very fluid, structure already present in the egg, and had concluded that the future individual had acquired its form at that stage,[86] such a designation was obvious, but on further consideration misconceived.

The rise of the concept of preformation and its counterpart, preexistence,[87] is definitely coupled to the introduction of the mechanical philosophy in the life sciences. As preformation entailed the growth of preformed structure through the incorporation of additional matter, it offered a feasible explanation of ontogenesis, but the mechanical philosophy as a motive is certainly not applicable to Swammerdam's notions concerning preexistence. Recent scholarship has demonstrated that Swammerdam's remarks on this topic ought to be evaluated in light of his emphasis on order and regularity in nature. It is argued that Swammerdam's conception of "invisible but essential principles," from which the future individual develops, and which forestalled the direct intervention of God in generation, stemmed from a concern to maintain predetermined order in nature.[88] Certainly, Swammerdam's rare remarks on the developmental process are ambiguous and obscure. His main point with respect to generation was

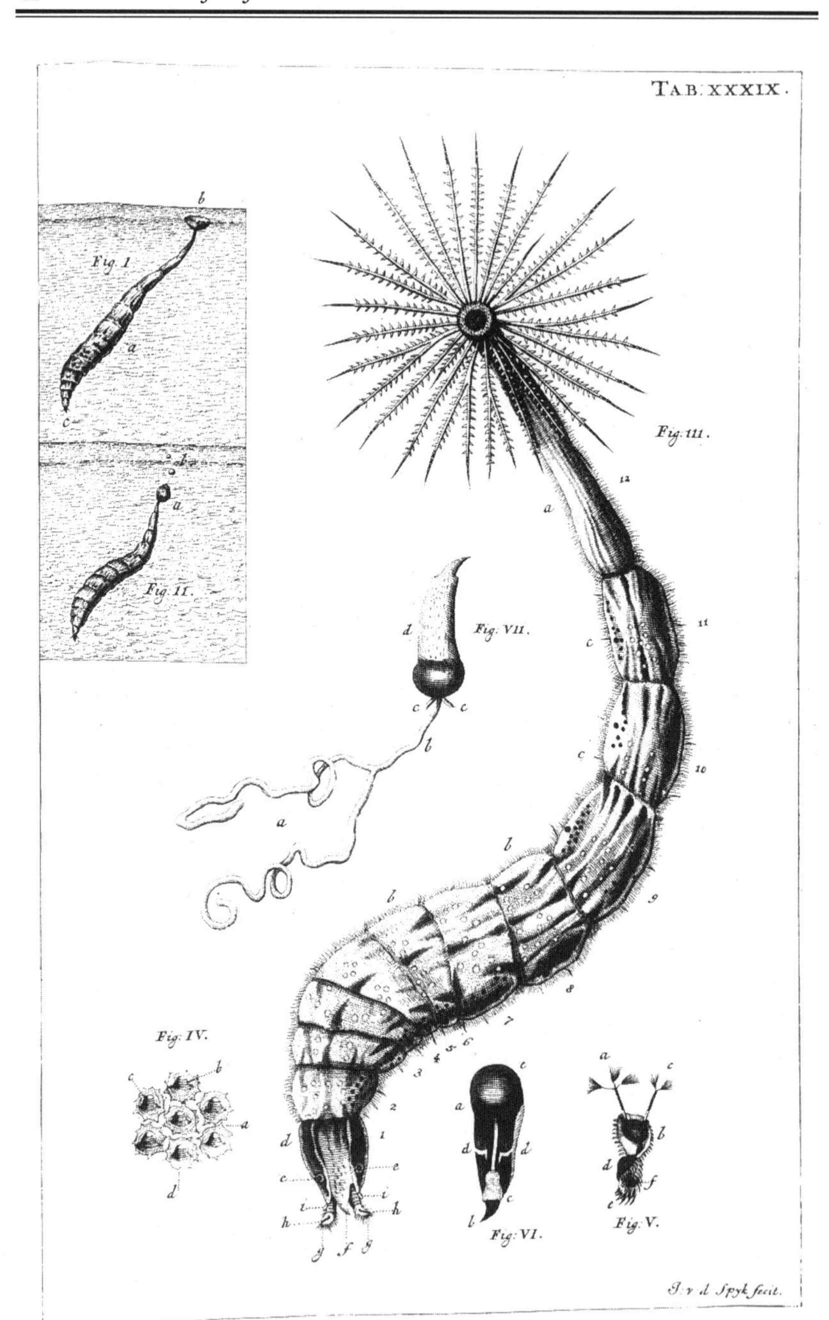

Fig. 10. Swammerdam's representation of the larva of the Asylus, table 34 in his *Biblia naturae* (1737–38).

that "there is absolutely no generation in the whole of nature, and not as generation, or growing of parts."[89]

This view implied that development was effected through growth of preexistent parts. Indeed, Swammerdam envisaged development initially as epigenesis, which included the growth, swelling, and budding of existing parts. As a result of subsequent research, epigenesis came to include the addition of novel parts and the loss and the rearrangement of parts. His investigations into a kind of fly called the Asylus (see figure 10), carried out in November 1677, disclosed some puzzling facts, namely, that the arrangement of the nervous system changed considerably, and that the gut of the larva disappeared while a new gut is formed in the imago. Swammerdam exclaimed in wonder, "It may be considered as a putting aside of the old parts, a new creation or a resurrection of the old body in a new."[90]

Swammerdam did not therefore imagine that a miniature of the adult animal was contained within the egg. He fully realized that the embryo differed considerably in appearance as compared to the adult, not only in insects, but in higher animals as well. With respect to the development of the tadpole on the second day he wrote that he only observed an accumulation of globules from which the animal appeared to take its origin.[91]

Swammerdam was in fact concerned with visible things, and his remarks on the invisible origins of the embryo may be regarded as loose speculations, fitting in nicely with some biblical problems or, alternatively, as rather ill-considered elaborations on the order prevailing in nature.[92] In neither case did he envisage a fully formed embryo within the egg, as contemporaries and eighteenth-century scholars proposed, but rather a scenario for the individual's future development, imprinted on the matter within the egg.

Order in nature, as perceived by Swammerdam, covered the unchanging sequence of stages in metamorphosis, the regular patterns in the life cycles of insects and their behavior, the similarity in the anatomy of all animals, including insects, the regular matching between a specific host and a specific parasite, and so on. Writing of the uniform master plan of animal anatomy he stated, "One can state with truth that GOD has created only one animal, which he has concealed and made distinct underneath an infinity of shapes, curves, convolutions, and stretchings of limbs: to which he has subjoined a different nature, way of life, and food."[93] This explicit statement concerning God's master plan obviously issued from the pen of an experienced anatomist struck by the similarities he had observed in the anatomies of a wide range of animals, whereas the statements concerning preexistence usually quoted from Swammerdam's writings were made by a man struggling with religious dogma.

Nehemiah Grew

Nehemiah Grew's incentive for studying the structure, development, and physiology of plants originated with the contemporary concern for, and investigation of, the manifold structural details of animal anatomy, a concern and investigation governed by the wish to resolve physiological questions. Reasoning by analogy, Grew concluded that the demonstrable phenomena of vegetation implied an equally detailed anatomy in plants. Grew's use of the microscope, for which he is famed in the history of science,[94] was only one of the various methods he applied in his researches on plants. Although by his own account the study of plants held his attention from 1664 up to at least 1677 (in which year he read the last of his lectures on botany to the Royal Society),[95] Grew's microscopic studies did not begin in earnest before 1672.

His first publication on plants, aptly entitled *The anatomy of vegetables begun,* was published in 1672 by order of the Royal Society[96] and contained only a few references to microscopic observations. These had been added to the already completed manuscript some time after Grew had met Hooke in London.[97] Not surprisingly, the observations that Grew did include were concerned with the cellular structure of plant material, as described earlier by Hooke. Grew duly acknowledged his debt to Hooke in the preface to his book, writing "after we had finished the whole composure, some observations made by that ingenious and learned person, Mr. Hooke, . . . and by him communicated to me, were super-added: likewise some others also microscopical, of my own, which his gave me the occasion of making."[98]

Once Grew had been appointed as a member of the Royal Society on 16 November 1671 he was, only a few months later, appointed as Curator for the Anatomy of Plants,[99] in which capacity he was to receive the amount of £50 a year, to be raised by the members. This office and the emoluments attached had been created especially for Grew, with the object of tempting him to continue his research in London instead of returning to his medical practice in Coventry.[100] On the occasion of the formal decision, the Council ordered that the society's microscope be placed at Grew's disposal.[101] Grew used the instrument to great advantage, as appears from the demonstrations and lectures with which he "entertained" his fellow members over the succeeding years. His last lecture containing microscopic data was delivered on 6 December 1677.[102] It was therefore only during a period of five years that Grew seriously studied the microscopic anatomy of plants.

Grew's association with the Royal Society was of crucial importance to his scientific career. First, the favorable reception of his original work on plants apparently stimulated him finally to obtain a formal degree in medicine. In the summer of 1671 he went to Leiden for a few weeks, matriculated there, and returned to London a medical doctor.[103] The fact that Grew chose to go to Leiden for the occasion was necessitated by reason of his religious background, as he came from a nonconformist family.[104]

Second, the Royal Society's decision to approve a scheme of financial support for Grew ensured that Grew's initial endeavors matured into a full-scale research program. Although the scheme for a salaried position did not materialize in the end, Grew did remain in London, supporting himself from emoluments received for lecturing to the members of the Royal Society, for example.[105] Eventually, however, he reverted completely to medical practice and quit research altogether.

The support of the Royal Society also helped Grew to overcome an initial drawback to his plans for continued research. As we have seen in the chapter on Malpighi, the latter's *Anatomes plantarum idea* reached the Royal Society on the very day that Grew presented his recently finished book to the fellows of the society. Naturally, Grew was rather discouraged by this development, but the encouraging attitude of the fellows made him decide to continue his research, accepting their reasoning that two could discover more than one.[106] In the succeeding years the Royal Society gave Grew suggestions for topics that they wanted investigated, such as whether the vessels in plants exhibit peristaltic movements.[107] More important, they ordered several of his lectures to be printed, which resulted in the publication of *An Idea of a Phytological History Propounded. Together With . . . the Anatomy of . . . Roots* (1673), *The Comparative Anatomy of Trunks* (1675), and *Experiments of Luctation* (1678).[108]

In 1677 the Royal Society appointed both Grew and Robert Hooke as its secretaries. From the records of the society[109] it is clear that Grew was rather busy for some time with the duties involved with this appointment, until Hooke more or less usurped these activities.[110] Grew then turned his attention to preparing an inventory of the possessions of the Royal Society, published in 1681 as *Musaeum regalis societatis.* Grew's active involvement with the society came to an end in 1684, when he resigned from the council.[111] The last of the projects he was engaged in as a member of the council was the preparation of the publication of his own collected works on plants, *The anatomy of plants,* which was published in 1682.[112]

Two of his fellow members of the Royal Society, Robert Hooke and Robert Boyle, exerted a particularly constructive influence over Grew's researches. Hooke had been engaged on microscopic research sporadically for over six years when Grew first came into contact with him through the Royal Society, and he had carried out some research on the problem of whether there are valves in the vessels of plants.[113] Thus, on the same day that the society decided to publish Grew's book, Hooke presented an experiment to the meeting concerning the transportation of mercury through the stem of a plant, and he discussed the "congeries of bladders, which have no visible communication."[114] Grew's subsequent researches on plant structure and his speculations on the life processes in plants started from the groundwork laid out by Hooke, but the scope of his work went far beyond anything that Hooke had attempted.

Boyle's influence on Grew's thinking concerning the nature of matter is properly acknowledged in the preface to *The anatomy of leaves, flowers, fruits and seeds,* which forms the fourth book of *The anatomy of plants.* Grew dedicated that part to Boyle, writing "in Discourse upon this Subject (= the *Reason* and *Scope* of Nature), *You* have been pleased frequently to insist, That I should by no means omit, to give likewise, some Examples of the *Mechanisme* of *Nature* in all other *Parts.* The Performance whereof therefore, . . . is to be looked upon, as a Due to the Authority which Your Judgment hath over me."[115] Grew's speculations and experiments concerning the chemical composition of plants closely followed Boyle's theories,[116] and the number and range of his experiments on vegetable and animal substances matched Boyle's output.

By the time Grew started to study the anatomy and physiology of plants in the mid-1660s, such an undertaking was a novelty. Until then plants had mainly been studied for their medicinal virtues. As a result botanical books, besides trying to bring comprehensible order into the rapidly growing number of known plants, were mainly devoted to the medicinal virtues of the plants. Grew, in his own account, conceived the idea of studying plants when he read of recent investigations into the bodies of animals, and the curious structures detected therein. Considering that both plants and animals "came at first out of the same Hand, and were therefore the Contrivances of the same Wisdom" he thought it worthwhile to investigate plants in order to reveal the structures of these organisms.[117] He was fully aware that in so doing he was about to tread on virgin ground. However, he expected that the study would be both intellectually rewarding and spiritually edifying. With respect to intellectual satisfaction he wrote, "Who would not say, That it were exceeding pleasant to know what we See: and not more delightful, to one who has *Eyes,* to discern that all is very fine; than to another who has *Reason,* to understand *how.*"[118]

It was evident to Grew in regards to the uplifting character of the study of nature that this particular study, which revealed the orderliness and constancy of structure in creatures, demonstrated the wisdom of God. Indeed, he wrote, "The higher we rise in the true Knowledge and due Contemplation of *This;* the nearer we come to the *Divine Author* hereof." In his books on botany Grew frequently expressed his view that, through scientific investigation, the wisdom of God becomes evident to man. Indeed, through science man comes to realize that "it is the Demonstration of *Divine Wisdome,* that the parts of *Nature* are so harmoniously contrived and set together."[119] For Grew religion followed naturally from science. His last published work, the *Cosmologia sacra* of 1701, was written with the object of demonstrating "that Religion is so far from being inconsistent with Philosophy, as to be the highest point of it."[120]

Grew professed himself to be wholeheartedly committed to the official philosophy of the Royal Society, which he characterized as an interdependent combination of reasoning, experimentation, and observation.[121] As a true Baconian Grew did not hesitate to accumulate a multitude of facts and, moreover, to present them all to his fellow members, although he may have felt some hesitation in so doing since he wrote, rather apologetically and by way of introduction to a presentation of his numerous experiments on organic substances, "No less a number would have answered the design of an *Universal Survey;* which, though less pleasing, proves the more instructive in the end: not being like angling with a single Hook; but like casting a Net against a shole: with assurance of drawing up something. Besides the advantage of *comparing many together;* which being thus joyned, do oftentimes, like *Figures,* signifie ten times more, then standing alone, they would have done."[122]

Throughout Grew's writings his observance of the Baconian principles of scientific endeavor is apparent. Thus the Baconian premise that "Natural and Artificial Mixture are the same"[123] lies at the basis of his chemical experiments and of his subsequent propositions concerning the chemical composition of plant materials. Grew also felt his exertions to be part of a much grander design, reaching back to the "Ancients" and stretching into the future, that is, the long and gradual journey that leads to knowledge of nature.[124]

In his *Idea of a Phytological History* (1673) Grew delineated an exhaustive research program concerning the structure, nature, and life processes of plants. He included a wide range of problems in this program: growth, nutrition, the form and dimensions of all the plant parts, development, propagation. Problems such as "Why do roots descend and stems ascend," "Why do plants that are structured from identical common parts, grow in the same soil with the same sun burning on them, the same rain pouring on them, have such different natures and faculties" are typical examples among the host of others that Grew thought should be investigated. The ultimate goal of the project was the elucidation of the "nature of vegetation."

To that end Grew proposed the use of five general methods, or "means" as he called them. These were: a survey of the external parts (morphology), the disposition of the original parts (anatomy), analysis of the contents (saps, etc.), analysis of the "principles" (i.e., the chemical constituents such as salt or water) of the organic parts, and last, an examination of the raw materials from which the plant grows. The investigations within each "general means" should cover a wide range of plants. The presentation of the results of these studies could, according to Grew, best follow the way of nature. That is, "The best Method of Delivery, for clear Discourse, can be but one, according to that of *Nature,* from the *Seed* forward to the *Seed.*"[125] In between, the root, stem, leaves, flowers, and fruits were to be dealt with. Grew was well aware of the magnitude of the rigorous and methodical program he had outlined. But he did not intend the program of his *Idea* solely as a guide for his own future research. On the contrary, he envisaged that some of the approaches he had indicated would appeal to other scholars, so that in the end the complete program would be executed through the joint efforts of a number of men.[126] Grew himself concluded his researches on plants by the end of 1677, five years after he had started on them. Indeed, his initial estimate of the time needed for the research he had envisaged had been about five years.[127] By that time Grew had already become involved with another scheme, the comparative anatomy of animals, with which the Royal Society was well pleased.[128] Yet this project was eventually abandoned, and Grew presented only a few lectures on this subject, while only one tract on animal anatomy was completed and published.[129]

Grew's speculations on the life processes in plants were determined, like those of many of his contemporaries, by extending a fundamental analogy from animals to plants.[130] In animals the various vital functions were known to be related to specific structures. Therefore, reasoning by analogy, the structures observed in plants had to serve specific functions. His observations established, wrote Grew, that "a *Plant,* as well as an *Animal,* is composed of several *Organic Parts;* some whereof may be called its *Bowels.* That every *Plant* hath *Bowels* of divers kinds, conteining divers kinds of *Liquors.* That even a *Plant* lives partly upon *Aer;* for the reception whereof, it hath those Parts which are answerable to *Lungs.*"[131]

Consequently Grew attempted to relate all the structures he had observed to their role within a comprehensive scheme of vegetation. This scheme accounted for respiration, nutrition, and generation, and involved a complicated circulation of air and nutritive saps through the plant. Chemical affinity and mechanical forces such as pressure and capillarity played major roles as causal agents within that scheme.[132]

Apart from this fundamental point, Grew's analogy between animals and plants was mainly limited to structural detail, in that he was apt to name plant structures in accordance with the similarly built structures of animals. In addition, he referred on occasion to animal equivalents by way of confirming his speculations. In one such case Grew argued for various kinds of air vessels in plants, referring to the differences in the organ of respiration in land animals and fishes.[133] Grew also thought that the study of plants "may frequently conduct our minds to the consideration of the *State* of *Animals;* as whether there are not divers material Agreements betwixt them both; and what they are. *Wherein* also they may considerably differ. . . . And *besides,* not only to compare what is already known of both; but also, what may be observed in the *one,* to suggest and facilitate the finding out of what may yet be unobserved in the *other.*"[134]

However, Grew's own venture into animal anatomy was a rather abortive one, so that the above surmise was not put to the test. Grew typically used domestic analogies when marveling about the intricate construction of plant matter.[135] His ideas on this subject were founded on an analogy with fabrics, lace, and basketry. The "Warp and Woof" of the fibers of which the plant was constructed was a pleasing notion to which Grew frequently referred.

Grew's conception of nature derived from two sources: his microscopic research and the corpuscular philosophy of matter. Grew extended the conclusions he had derived from his microscopic observations from the vegetable to the animal world, writing in his *Cosmologia sacra,* "And herewith [the structure of vegetable parts] there is a great agreement in the Structure of the Organs of an Animal. That the Muscules, Membranes, and Skin, are composed of Fibers, is well known to Anatomists. And I add, here . . . that Cartilages, and Bones themselves, originally, and all the *Viscera* are composed of Fibers. And it is probable, that these Fibers are or once were also Hollow: for the conveyance either of a Liquor or an Aerial Spirit."[136] Thus he envisaged living matter as constructed from infinitely small, presumably hollow, fibers.

A part of Grew's research was concerned with demonstrating the nature of organic matter. He was, in the wake of his fellow member of the Royal Society Robert Boyle, an advocate of the corpuscular philosophy of matter. In his lecture on the "Nature, Causes, and Power of Mixture" Grew explained, at length, the various ways in which the several kinds of indivisible atoms could be mixed together to form diverse kinds of materials and the various manners in which the atoms might cling together. In this lecture he substantiated his thesis that all matter, including organic bodies, is formed by mixture by describing some experiments in which he imitated the creation of animal, vegetable, and mineral bodies by mixing several substances and applying the appropriate operative "causes" (such as agitation or solution).[137]

Conversely, he held that the composition of materials may be studied by mixing them with specific substances and observing the reaction of the mixture. Thus, the mixture served as "a Key to let us easily into the knowledge of the Nature of Bodies."[138] In a further study Grew observed the results of "calcining them, or as it were, by mixing them with the Fire." In this case he studied mostly plant materials. In yet a further elaboration of his point, Grew studied the effect of "*mixing* them with *Aer* or exposing them to it."[139]

In Grew's opinion the chemical composition of the parts determined their form. Central to this view was the idea that like attracts like, and that a particle of a certain substance would therefore preferably adhere to another particle of the same kind. The shape of the respective particles would determine the form of the parts they made up, because an elongated particle would cause a growth in the length. "Thus," he wrote, "doth Nature every where *geometrize.* For what She appears in Her *Works,* She must needs be also in their *Causes.*"[140] The predominant principle in a certain structure was, as it were, "the *Mold,* about which, the other more passive *Principles* gathering themselves, they all consort and fashion to it."[141] Thus Grew explained the longitudinal form of the vessels by their high proportion of alkaline salts, which of themselves shoot out lengthwise. In the case of the "Aer-vessels" (spiral vessels in modern terminology), the interaction between the alkaline salt and the "crooked particles of the air" caused the spiral thickening on the wall of a longitudinal element: "The said *crooked* Particles of the *Aer,* first *shooting* and *setting* together, as the *Mold,* the other *Principle* cling and *fix* conformably round about them. So that, as by force of the *Saline Principles,* the rest of them are made to *shoot* out in *Long continued Fibers;* so by force of the *Aerial,* those *Fibers* are still disposed into *Spiral Lines,* thus making up the *Aer-Vessels.*"[142]

The chemical composition of the parts was, according to Grew, the first cause of their shape (and also of the other properties), whereas he regarded the "visible *Mechanism* of the *Parts*" (that is, the arrangement of the various parts of the plant) as "a secondary *Order* of *Causes;* which serve rather to carry on and improve, that which *Nature* hath once begun."[143] Therefore the properties of the various structures in plants were determined both by their chemical composition and the position occupied within the anatomy of the plant.

Antoni van Leeuwenhoek

Leeuwenhoek did not begin his microscopic research until he was nearly forty years of age, at which time he was employed by the city of Delft in several capacities. He had no academic training; in fact he had had no schooling at all after the age of sixteen and spoke, wrote, and read only his native Dutch. It has been argued that these circumstances constituted serious drawbacks, because they meant that Leeuwenhoek had few acquaintances in the world of the learned who might have helped him to direct his researches more profitably, and with whom he might have discussed speculations founded on his observations.[144] Moreover, it meant that he was not conversant with the scientific views and theories concerning the "oeconomy" of plants and animals current in his time, which could have stimulated his theoretical thinking. However, the lack of such contacts implied that he did not have to discuss his work with people criticizing his work, and that was perhaps for the best, since he "did not feel inclined to stand blame or refutation from others."[145] However, during the half century that he was actively involved with microscopic research he became familiar with much contemporary science, through books written in or translated into Dutch, and through helpful friends and correspondents. Then, however, with characteristic hauteur he only bothered to discuss those that agreed with his own interests and theories.[146]

On the other hand, some have argued, Leeuwenhoek's rather isolated scientific position also had its positive effects, particularly on his theoretical thinking.[147] He was not hampered by traditional views that might not have agreed with his research, and he could therefore interpret his results in an unbiased fashion. Be that as it may, Leeuwenhoek became a very famous man, visited by an almost continuous stream of people, among them well-known scientists such as Christiaan Huygens, Jan Swammerdam, and Herman Boerhaave. Moreover, the town of Delft enjoyed the sight of a disproportionate number of royal visitors, among them Tsar Peter I and Queen Mary of England. Leeuwenhoek did not actually meet the latter, being out of town on that particular day.[148] Although visits between well-known scientists and the social elite were customary at the time, the fact that Leeuwenhoek was singled out by an overwhelming number of people is evidence of the measure in which his researches captured the imagination of his contemporaries. Of the leading microscopists of his age Leeuwenhoek undoubtedly attracted the most attention, and this has continued into the present. Ever since the establishment of the cell theory by the middle of the nineteenth century lent a new perspective on the spermatozoa and the unicellular organisms, Leeuwenhoek's microscopic researches, particularly those concerning the entities mentioned above, have been the object of historical analysis.[149]

Leeuwenhoek's work, without discounting the various important discoveries he made, has nevertheless been regarded in controversial terms both by contemporary scientists and scholars, as well as by historians of science. On the one hand, the rather muddled presentation of his work has tempted some to regard this particular microscopist as a true amateur, whose researches could hardly qualify as science.[150] On the other, careful scrutiny of Leeuwenhoek's letters has convinced a great many others that, even if the results were not presented in a very consistent manner, his method of working does indeed conform to the highest standards that might be imposed on a researcher at the time.[151] In regard to the presentation of his work, it has been pointed out that Leeuwenhoek deliberately fashioned his letters so as to appear to be reports of current research, including the various diversions from his main line of enquiry.[152] As to the scientific character of his output, a conflict was—and still is—perceived as existing between Leeuwenhoek's undisputed merits as an observer and his interpretation of the data. The latter has been subjected to doubt and even ridicule.[153] The influential Constantijn Huygens the elder pointed out that although Leeuwenhoek was unlearned both in sciences and languages, he was an exceedingly curious and industrious man.[154] On the strength of these qualities of character he expected great achievements from Leeuwenhoek, and he furthered the latter's career in science. Yet a good many of Leeuwenhoek's contemporaries, among them Thomas Molyneux, Nicolaas Hartsoeker, and to some extent Christiaan Huygens, would have agreed with Jan Swammerdam, who deplored that Leeuwenhoek was biased and could not discourse in a proper scholarly way.[155] Nevertheless, many a person reading one of Leeuwenhoek's letters containing the painstaking details of some of his multifarious observations would have agreed with Leibniz that they preferred Leeuwenhoek's observations to a Cartesian's deliberations.[156]

Leeuwenhoek's research, which covered a period of nearly fifty years, was communicated to the scientific community largely via the *Philosophical Transactions.* More than half his letters were addressed to the Royal Society. His townsman Regnier de Graaf had introduced him to the society in 1673.[157] Leeuwenhoek's work was received with enthusiasm by the members of that body, which counted among its members not only three of the outstanding microscopists whose work we have discussed in the preceding sections—Hooke, Malpighi, and Grew—but also several other men who at times actively employed the microscope, for example, Christopher Wren, Henry Power, Edmund King, Edward Tyson, and Martin Lister. It is no wonder that Leeuwenhoek's letters were enthusiastically received and discussed. Sometimes a letter led to attempts to verify his observations.[158] Moreover, the members of the Royal Society, as was their wont within their own circle, suggested to Leeuwenhoek topics that he might investigate. This interaction might partly explain why a letter by Leeuwenhoek often broached one subject after another, without any apparent relation between the two.[159] However, he deliberately exaggerated this effect, as he thought that in this way his letters would appeal more to the taste of the Royal Society.[160] The society did indeed look favorably on Leeuwenhoek's research and appointed him a member in 1680.[161] It is therefore clear that the style of Leeuwenhoek's research and the presentation thereof were strongly influenced by the Baconian character of the Royal Society.

Leeuwenhoek's letters, both those to the Royal Society and to various persons of renown, were also published in separate volumes, in both Dutch and Latin versions, but Leeuwenhoek made no attempt to organize their contents into a more coherent form by, for instance, uniting the passages dealing with the same topics, dispersed over several letters. This may have been partly due to the fact that it was not the author of the letters who took the initiative for this publication but a publisher,[162] who presumably expected a good sale of the letters written by the now famous "Delftenaar," and therefore ventured the publication of some letters in 1684. In later years Leeuwenhoek collaborated in the publication of his collected letters, but the pattern, that is, the original letters in their entirety, had been firmly established by then.

Particularly relevant to the present enquiry is the fact that Leeuwenhoek, having had no prior schooling in science and therefore having no predetermined framework with which to interpret his observations, may be assumed to have derived his conclusions directly from his microscopic observations. His accomplishments therefore offer an opportunity to study what the microscope could offer a mind bent on elucidating the workings of nature in living bodies.

One of the most conspicuous characteristics of Leeuwenhoek's work was that he studied a subject in fits and starts, sometimes over a great span of time, and consequently broached that subject in a larger or smaller number of his letters at a given time. This way of working has been called his "concentric method."[163] He thought that he would "come closer to the truth the more often he concentrated his investigations on one and the same thing at different moments."[164] His research on the spermatozoa offers a perfect example of this method.

Having discovered the animalcules in semen in 1677, and almost immediately convinced that the sperm constituted the material beginning of the future individual, Leeuwenhoek undertook during the subsequent eight years a series of investigations to elucidate and strengthen this thesis. Apart from establishing the occurrence of the animalcules in a number of different kinds of animals ranging from the louse to man, he established that the animalcules are formed in the testicles (1679), that the animalcules enter the uterus after copulation (1685), and that the genitals are specifically shaped so as to ensure that the animalcules enter the uterus (1685).

The discovery of the animalcules in sperm followed closely on a recent reappraisal of the function of the ovaries in generation. The careful research carried out by Regnier de Graaf had resulted in the thesis that the follicles of the ovaries contain the mammalian egg (although de Graaf did not in fact see this egg). The mechanism of generation was thereupon explained by the notion that the egg constituted the material beginning of the future being, while the sperm, through immaterial influences, caused the onset of the development of the egg. Leeuwenhoek's thesis therefore cut right across a recently established consensus. Consequently, he tried to refute de Graaf's thesis, concurrently with his research on the spermatozoa. He was so convinced of the validity of his own theory that he did not study de Graaf's writings very closely, and he consequently began his campaign with the mistaken assumption that the follicles themselves are these "eggs." The follicles, he found (1678, 1683), cannot be separated from the ovaries without causing serious damage, and their diameter is too large for them to pass through the oviduct on their way to the uterus. Therefore, the idea that the ovaries have anything to do with generation is utter nonsense, he decided.

Leeuwenhoek advocated the theory that the animalcules in the sperm are in fact preformed organisms, although he never claimed that he could distinguish the parts of that being within the animalcule. Repeatedly he wrote to the effect that "the Human Creature is enveloped in an Animalcule from the Male sperm, but it is incredible to me that human intelligence will penetrate this great mystery so deeply that by accident or upon dissection of an Animalcule from the Male sperm we shall come to see a whole Human being."[165]

His view that the animalcules were already completely structured organisms was reinforced by his research on plants. In the seed of the plant he discerned the plant embryo, and the embryo he saw was pervaded by small vessels. If a plant embryo contained these minute vessels, which corresponded to the vessels of the adult herb or tree, so (by analogy) the animalcules in the sperm were endowed with the rudiments of the anatomy of the adult organism as well. Over the years he also studied many other salient aspects of the spermatozoa, such as their lifespan outside the body and the speed with which they penetrate the uterus. He also tried to recover the animalcules in the eggs of birds, frogs, and insects, but failed.

These are the essential features of Leeuwenhoek's research on the animalcules in the semen, and his subsequent theorizing on the subject. By 1685 he had thoroughly investigated the various implications of, and objections to, his original premise concerning the role of the spermatozoa in generation. Over the ensuing four decades he frequently affirmed the various particulars of his theory and investigated several aspects anew, usually in the course of checking the facts in a hitherto unexplored subject.

An analysis of the development of Leeuwenhoek's investigations on various other important topics of his research, such as the capillaries of the blood, the woody parts of the plant stem, the muscle, and the unicellular organisms (on the last-mentioned topics see chapter 5), reveals a similar pattern. His ideas were formed shortly after he had launched on a particular topic, but over the years he returned to the same subject to study the same problem investigating another species. He therefore arrived at a definite view on his subjects even in the first stages of his inquiries. Although repetition of substantially the same research sometimes yielded new structural particulars or resolution of hitherto obscure details, this rarely led to a shift in his interpretations. In sum, Leeuwenhoek's "concentric method," which he himself expected to lead him nearer the truth, did indeed lead to greater anatomical accuracy, but seldom yielded results that would not fit into his theories conceived earlier.

Leeuwenhoek, it may be concluded, was particularly concerned with structure. It was his main concern to reveal the hidden structure of living matter. This being so, and since he was at the same time engaged with charting a virtually unknown terrain, Leeuwenhoek was confronted with a problem of terminology that he was unable to solve. In fact, it is a serious drawback of Leeuwenhoek's presentation of his research and one of the main reasons why his work at first sight strikes one as being muddled that he used a very limited array of technical terms. This can hardly be surprising, because so many of the structures that he saw had not been seen or even suspected before the microscope came into use. Moreover, Leeuwenhoek discovered a good many structures and even complete organisms earlier than any of his fellow microscopists, so that he had to use either conventional and everyday descriptive terms, or develop a new set of terms himself. He adopted the former course, which can be construed as the outcome of his lack of formal training in science.[166] Lacking a feeling for the need of discriminative terms, he used the same denominator for entirely different things, solely on the ground of their overall form. For example, throughout his letters he referred to the spermatozoa, a host of different infusoria, the smaller water insects, and many other insects as "dierken" or "diertgen" (animalcule); the word "globule" may denote anything from a drop of oil to the contents of the unicellular organisms, the component of a red blood particle, or the universal structural element that Leeuwenhoek envisaged up to 1680.

Another characteristic feature of Leeuwenhoek's work was his quantitative approach. This was no doubt stimulated by his professional activities (Leeuwenhoek being among other things an inspector of weights and measures in the town of Delft). He measured the things he saw by a kind of standard measure taken from natural objects that he could put under the microscope alongside his objects, for example, a grain of sand and a piece of hair from his beard or wig. As far as some biographers were able to determine,[167] these measures do give a fairly reliable result for the order of magnitude in the things he described. Leeuwenhoek's interest in mass and numbers also showed itself prominently in his calculations concerning the amount of animalcules in the sperm and in a drop of water, the diameter of the vessels in an infusorian or spermatozoon, the relative sizes between several kinds of organisms, but also in a quantitative analysis of reproduction.[168] Most of these calculations served to demonstrate how infinitely small the structures of the living organisms were, and that they were therefore ultimately intended as an illustration of the majesty and wisdom of God.

Leeuwenhoek made frequent use of analogies to arrive at a clear understanding of his observations. A case in point is his study of the plant embryo, which he thought to be directly analogous to the spermatozoon. As he could easily discover the particulars of a miniature plant in the seed of plants, he was assured that an equally structured being was lodged within the spermatozoon, even though he could not resolve any structure within that animalcule. In the same way he extended his findings concerning the annual rings in plant tissue to various other circular patterns, for instance, to the scales of fishes, in which he also interpreted the circular pattern as the result of periodical growth.[169]

Besides drawing analogies within the realm of organic matter, he also relied on analogies with mechanical models. For instance, he substantiated his view of the transportation of fluid through the leaves of a plant with a reference to a model of adjoining spheres of clay.[170] When the outermost of the spheres was wetted the fluid gradually spread through them all; in the same manner the plant vessels soaked up water from neighboring vessels. Another case in point is the experiment in which he established that air cannot pass through the intestine wall, while water can. The experiment was performed with an inflated and dried bladder to which was tied a water-filled length of gut.[171] Leeuwenhoek observed that the water soaked from the gut onto the bladder, but that no air escaped from the bladder.

A much-praised quality of Leeuwenhoek's attitude to scientific endeavor was that he did not hesitate to retract his ideas once he realized that he had made a mistake.[172] He wrote as much to his correspondents in the Royal Society[173] and did openly admit mistakes in several instances.[174] It should be noted, however, that on these occasions he merely offered corrections of his own earlier observations.

Leeuwenhoek was loath to consider seriously, however, an opponent's objections in a dispute. In a reply to a letter from George Garden, containing a serious and scholarly appraisal of the role of the mammalian ovaries in generation,[175] he pronounced, before even considering Garden's arguments, that "I am still of the opinion that these (viz., theories concerning the function of the mammalian ovaries in generation) are mere figments, because in former days people have not been able to invent any better theses concerning generation."[176]

Garden had put forward the view that the animalcules can only develop when lodged in a proper nidus, which he situated in the cicatricula of a fowl's egg. He equated the mammalian egg with the cicatricula, which is therefore an egg devoid of any nourishment for the developing embryo. He advanced various arguments in support of such a view. One of these was that some time elapsed between conception and the moment that the developing embryo started to send forth the minute vessels that will eventually form the umbilical cord. Consequently the cicatricula, which sustains the embryo until it becomes attached to the wall of the uterus, must originate from outside the uterus, presumably the ovaries.

Leeuwenhoek replied that he had "not made any accurate investigation" of the early embryo's development. Nevertheless he pointed out that, even when no vessels were seen to emanate from the embryo, it did not necessarily mean that these were not present, but only that they were too small to be seen as yet. Besides, such vessels were presumably very fragile and would therefore break easily. Moreover, he argued, as the embryo has started to grow and cannot have been nourished from an egg, it must have received food from the uterus, even if no vessels can be detected. Leeuwenhoek, therefore, appealed to the presence of imperceptible structures to invalidate his opponent's argument, thus creating a stalemate position because his claim defied confirmation.

Other cases that indicate an aloof, self-centered stance are Leeuwenhoek's consistently wrong interpretation of the trachea in insects, and his rejection of boxlike structures in plants. Whereas Swammerdam and Malpighi had recognized the trachea as integral parts of the respiratory organ, Leeuwenhoek regarded them as blood vessels. He never appeared to realize that his view was at variance with that of Swammerdam and Malpighi, although he was conversant with the research of both men. In the case of the structure of vegetable tissues, he completely rejected Grew's notion that the medullary rays were composed of separate cells similar to those observed in the pith (a view shared by Malpighi and Hooke) without so much as an argument.[177] Moreover, when a discussion arose concerning the notion of an animal physiology based on the course, diameter, and other aspects of the blood vessels, which originated in his native country and for which his own research provided important grounds, he did not enter into this discussion.

Although Leeuwenhoek had no formal schooling, it is nevertheless obvious from the contents of his first letters that he was already conversant with the Cartesian conception of nature at the time he embarked on his career in science. In the course of the first two years of his communication with the learned world he dealt with such topics as the transport of fluid through plants,[178] the growth of animal tissues and plants,[179] the conduction of stimuli through nerves,[180] and the relation between the arrangement of the constituent parts and the characteristics of tissues,[181] in a perfectly Cartesian fashion. That is to say, he allocated a fundamental role to colliding corpuscles in these respects. Equally telling about his partiality is perhaps his early excursion into a problem of physics. Writing about his experiments on the compression of air he relied on the "first and subtilest stuff of the Air," a concept frequently encountered in Descartes's writings, to account for his results.[182] In subsequent letters Leeuwenhoek now and then endorsed generally supported mechanical notions, such as the mechanical origin of taste, in which taste is caused by the shape of the crystals of sugar, salt, and so on.[183]

In keeping with the Cartesian conception of nature, Leeuwenhoek favored an idiosyncratic corpuscular theory of matter.[184] In the early years of his investigations he discerned globules in nearly all of the objects he studied. Consequently he formed the idea that all matter was composed of globular units. Notwithstanding the serious doubts of others, such as Christiaan Huygens, he developed this theory to a certain extent.[185] He thought of these units as rather malleable, and assumed that they could change form as easily as could a bladder filled with water. Therefore they remained joined tightly together, leaving no open spaces between them when densely packed. These globules were not equivalent to the smallest extant particles but were, rather, composite units, certainly in the case of organic matter. Leeuwenhoek thought of these globular units as composed of smaller particles and of their formation and maintenance as a continuous process; he imagined that the smallest globules were continually supplied with fresh nourishment, that new particles were continuously created, and in the meantime others were expelled.[186]

However, within a few years Leeuwenhoek came to realize that he had founded his globule theory on accidental circumstances. When he repeated an investigation of the fabric of teeth, which he had formerly asserted were built from numerous globules, he now found that he had "erred with regard to the globules, which we imagined to see so distinctly that we could not but take them for globules; especially when I used much stronger glasses, with greater magnification, for then I could see quite clearly and distinctly that the whole tooth was made up of very small transparent pipes."[187]

The conclusion was plain: what he had formerly thought of as globules were in fact cross sections through small pipes. From then on he observed ever more longitudinal elements in living materials. Consequently his speculations concerning the organization of living matter were directed onto another track. Whereas he had formerly envisaged a uniform unit, he now assumed a variety of structural elements whose form corresponded with the main characteristic of the composite structure. The blood corpuscles, for instance, he thought of as composed of six globules, and each of these globules compiled from another six globules of a smaller order of magnitude (see figure 11).[188] Muscle fibers were constructed from yet smaller fibers and crystals from yet smaller crystal-like entities.[189] Even more importantly, he came to realize that just beyond the reach of his microscope his specimens were still structured from similar elements. The ultimate structural elements were therefore invisible, and he doubted whether they would ever be resolved. Indeed, he wrote that the particles of which water was constituted were too small to be seen, even with the microscope, which did not mean, however, that he had no definite ideas as to the form of these particular particles.[190] Therefore the impression is given that Leeuwenhoek came to know of the essentials of Descartes's thinking on matter early in the 1670s, found much to his liking therein, and proceeded to apply Descartes's notions to his own as yet unformed investigations and conclusions. However, it appears that Leeuwenhoek was not really well informed about the finer details of Descartes's views. Thus as soon as he could no longer reconcile simple Cartesian notions with his microscopic observations, he simply quietly dropped the former from his writings.

In the course of his long career, Leeuwenhoek's views on the organization of living matter were gradually adjusted to his observations. He observed minute structure everywhere, and studying the same object with a better lens, he detected yet more detail. It is no wonder that he assumed the animalcules both in the sperm and in watery environments to have an intricate structure, comparable to that of much larger animals; indeed, he thought the parts contained within these minute organisms to be no fewer in number than those in the bodies of man.[191]

This supposition derived to a large extent also from his conviction that nature acts in a uniform way everywhere. "The Author of Nature," he had

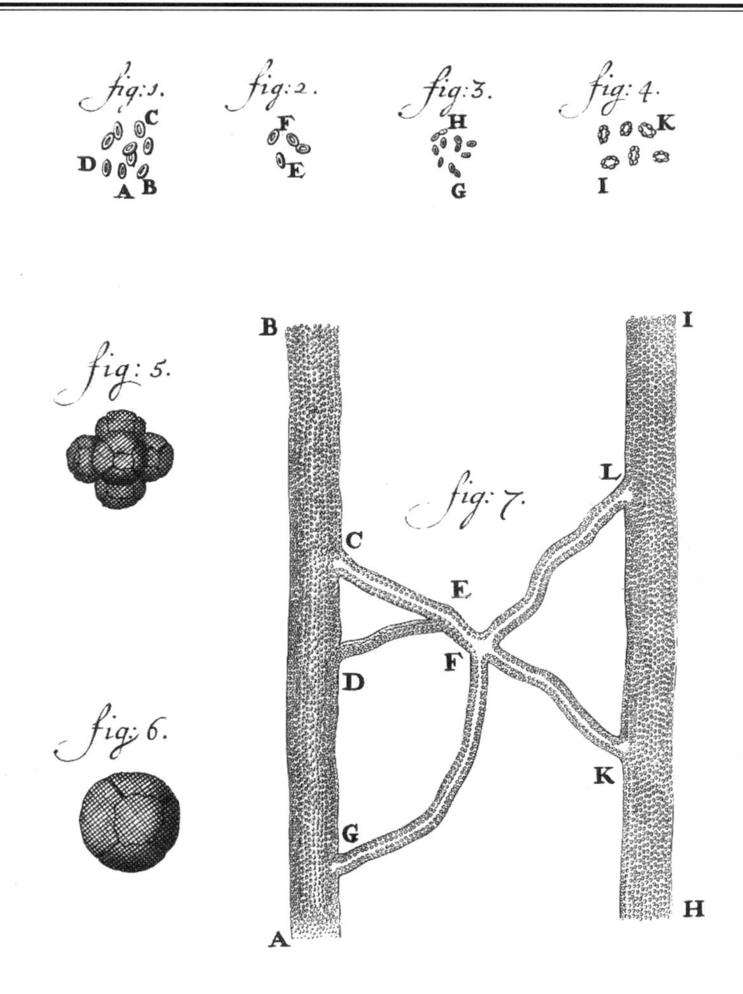

Fig. 11. Leeuwenhoek's representation of the capillaries and (in figs. 5 and 6) his wax model of the structure of the red blood cells, from the letter dated 9 July 1700, in his *Sevende vervolg der brieven* (1702).

observed, usually employed the same plan of construction in a wide variety of creatures.[192] Insects, it had been discovered, have organs like mammals, and by analogy Leeuwenhoek thought that even minuter organisms must also have comparable organs. He argued, "When I see animals living that are more than a hundred times smaller and am unable to observe any legs at all, I still conclude from their structure and the movements of their bodies that they do have legs and therefore legs in proportion to their bodies . . . and these legs, besides having the instruments for movement, must be provided with vessels to carry food."[193] He did not doubt the truth of this parallel and calculated without qualms the diameter of a blood vessel in such a leg.[194]

The notion of uniformity in nature was in fact the most important postulate in Leeuwenhoek's reasoning. This postulate warranted the application of his separate observations throughout the realms of nature. The ultimate outcome of the notion of uniformity in nature as regards the organization of living matter was the doctrine of the preformed embryo. The same postulate underlies his vehement rejection of spontaneous generation.

Leeuwenhoek spurned spontaneous generation for several reasons and found proof particularly in the elucidation of the life cycles of insects and the anatomy of the sexual apparatus and the presence of spermatozoa.[195] Within the context of this issue he studied a variety of common and more unusual insects such as the calander, weevil, grain moth, flea, and louse.[196] When the presence of spermatozoa in a species of mites was established it was to Leeuwenhoek's mind conclusively demonstrated that all animals, however small they may be, take their origin exclusively in procreation.[197]

In the course of his studies he discovered a number of obscure insects, such as the aphids. These particular organisms presented him with a difficult problem. He observed that the young, which contrary to expectation were born instead of being hatched, reproduced both before and after they had matured into winged adults. Moreover, no males were involved in the process.[198] Rather anxiously Leeuwenhoek concluded that aphids propagate without intercourse, which he feared to be grist to the mill of the proponents of spontaneous generation. He found a solution to this problem through the use of the principle of uniformity. Indeed, when pondering the implications of his doctrine of spermatozoa, he had calculated the total number of animalcules that could be discharged by a male of the species in the course of a year, and had concluded that such an amount could not possibly have been contained at any one time by the testicles. He therefore assumed that the animalcules were somehow generated within the testicles. He postulated a "seminal stuff" that was left behind by the departing spermatozoa and from which new spermatozoa were generated without intercourse, just as in the case of the aphids in fact. Thus, he felt, he had restored the aberrant generation of the aphids to his original scheme.[199]

In spite of his reliance on the principle of uniformity, where the appositeness of his theories throughout the realms of nature was concerned it did not affect his program for further research. A case in point is the structure of muscles: although he was already convinced that the flesh of all quadrupeds consists of fibers, he nevertheless decided to observe the muscles of several species of animals with the microscope.[200]

Leeuwenhoek was primarily interested in discovering the function of a structure, particularly as he could not imagine that provident nature had produced any redundant structures, except of course the ovaries, to which he denied any function, a denial he supported with a reference to the non-functional nipples on a man's chest.[201] In the course of his investigations concerning the infusoria and small aquatic animals he observed the water current caused by the "wheels" of a species of rotifer.[202] He was not satisfied about its function until he had actually observed how particles floating in the water were sucked toward the animal as a result of the swirl created by the wheels. Only then did he feel assured that this was the way these animals drew their food within reach.[203]

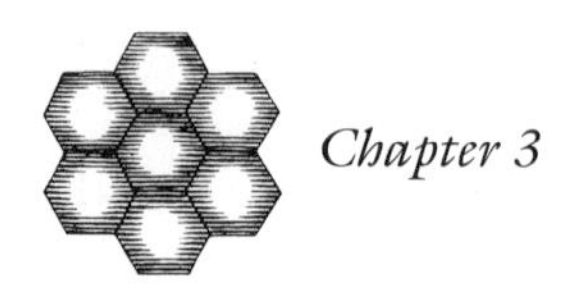

Chapter 3

The Substance of Living Matter

DURING THE 1660s the mechanical system of thinking took a firm grip on the life sciences, resulting especially in mechanical explanations for the various physiological processes. Some, among them Henry Power and Thomas Willis, took atoms as the starting point for their experiments, observations, and rationalizations, while many others, particularly in Italy, focused on the physical properties of the organs or the texture of the various tissues. Well-known examples of the mechanical explanation of physiological operations are Stensen's geometric theory of muscle action, and the explication of vision through the refractory properties of the eye lens and of secretion by the sievelike action of glands. However, the application of this system of thought to the phenomena of living bodies initially had greater appeal for physicists than for physicians. The heightened interest in the organic world from these quarters was to a large extent prompted by their exploration of the implications of the mechanistic hypotheses. The ultimate consequences thereof were debated specifically in connection with reproduction.[1] The mechanisms at work in the creation of a new organism are obviously crucially important when considering the question of whether living bodies function with or without the intervention of a "prime mover," or God. It was precisely this question that touched the core of the mechanistic philosophies.

Consequently, physicists such as Pierre Gassendi, hoping to catch a glimpse of the elusive particles of which matter was thought to be composed, examined various sorts of materials (including small beasts) with the microscope, and in so doing they became interested for a time in topics from the life sciences.[2] In fact, microscopic examinations did not provide direct empirical proof of the existence of particles of matter, but rather rendered support for the extrapolation from well-known facts to the level of corpuscles. Indeed, the microscope clearly demonstrated that, beyond the structural level distinguishable with the naked eye, ever more complicated structures could be resolved. A typical example of the way in which microscopic observations were used to substantiate such arguments is to be found in Walter Charleton's *Physiologia Epicuro-Gassendo-Charltoniana* of 1654. Reflecting on the magnitude of atoms he mentioned the microscopic image of a "handworm," which disclosed "an oval-head, and therin a mouth, or prominent snout, armed with an appendent proboscis, or trunk consisting of many villous filaments contorted into a cone, wherewith it perforates the skin, and sucks up the bloud of our hands; but also many thighs, leggs, feet, toes, laterally ranged on each side; many hairy tufts on the tail, and many asperities, protuberances, and rugosities in the skin."[3]

These observations were thereupon used to argue that the inside of the animal must surely have a complete anatomy so as to take care of nutrition, locomotion, and sensation. The minuteness of these parts (which Charleton had not seen, only deduced) was in its turn used as an argument for the minuteness of the atoms. Charleton called the use of such arguments the "Dioptrick of Reason" or the "Engyscope of the Mind."

Henry Power

Several writers, among them Robert Boyle, used microscopic evidence as a starting point for similar arguments, yet some investigators expected, or at least hoped, that the basic elements or atoms might actually be observed with the microscope. This expectation was widely shared at the time. Henry Power, who was a convinced corpuscularian and a microscopist as well,[4] expected the microscope to reveal the minute particles of matter in the future, writing, "And indeed if the Dioptricks further prevail, and that darling Art could but perform what the Theorists in Conical sections demonstrate, we might hope, ere long, to see the Magnetical Effluviums of the Loadstone, the Solary Atoms of light . . . the springy particles of Air, the constant and tumultuary motion of the Atoms of all fluid Bodies, and those infinite, insensible Corpuscles (which daily produce those prodigious (though common) effects amongst us)."[5] Even so, he reported faithfully that despite several efforts, he himself had been unable to catch a glimpse of the atoms.[6]

Henry Power's treatise on microscopic observations shows a close link between the application of the microscope and corpuscular philosophy. Power was a physician by profession, but diligently studied the physical sciences. He was an early member of the Royal Society, and although the activities of the society then included a relatively large amount of microscopy, Power did not take an active part, probably because he lived far away from London, in Yorkshire. Just before Hooke published his *Micrographia,* which surpassed Power's attempts to an immeasurable degree, the latter's *Experimental Philosophy* was published in 1664. This book contains, besides two essays on magnetism and the vacuum, a survey of Power's microscopic investigations. Power, like Hooke, was convinced that by means of recently invented and developed instruments, including the microscope, the fundamental mechanism of nature would be detected. Indeed, he wrote that the philosophers' finest theorizing would be but empty conjectures if they did not make the best use of the instruments now available.[7]

Power's *Experimental Philosophy* contains many of the arguments underpinning the mechanistic worldview and expectations concerning the usefulness of the microscope for scientific enquiry, which are also found in the works of such contemporaries as Gassendi, Charleton, and Boyle. Power's microscopic research stands over and above the short references to microscopic investigations by these scholars. He inspected a fair number of specimens, including several species of insects; vinegar eels; a snail; crystals of sugar, sand, and salt; quicksilver; part of a ribbon; the leaves, seeds, and pollen of various plants; the sparks produced by flint; and hair. The descriptive part of this treatise, which had no illustrations, save for two or three very crude woodcuts, is adequate, but as Power did not attempt to dissect any specimens, apart from cutting off their heads, his descriptions mainly concern the exterior of his specimens. He based five general conclusions on his observations.

First, he decided that all imperfect animals have a pulsating heart and a circulation of nutritive fluids. He had actually observed through their integuments that the interior of the insect body contains a pulsating heart. "In these pretty Engines," Power concluded, "are lodged all the perfections of the largest Animals; they have the same organs of body, multiplicity of parts, variety of motions, diversity of figures, severality of functions with those of the largest size: and that which augments the miracle is, that, all these in so narrow a room neither interfere nor impede one another in their operations."[8] The very smallness of insects, which yet possessed the elementary equipment for life, impressed him deeply, and he wondered about the size of the smallest living entity, which would still display all the functions associated with life.[9]

His second conclusion was that the nutritive liquid in imperfect animals is, because of the absence of some "higher heat," not turned into blood. Power substantiated this conclusion by pointing out that, in the early stages of the development of perfect animals, a transparent liquid circulates, as may be seen in the cicatricula of the chicken embryo before the heart is formed. Nevertheless, this conclusion is somewhat surprising in view of the fact that Power had also observed that the red color of blood apparently disappears as blood is sucked into a narrow capillary. He took this to mean that the blood within the small blood vessels of such organisms such as insects only appears to be colorless.

The examination of a snail led Power to his third conclusion, namely, that the animal spirits circulate through the body just like blood. He had observed that as the snail slithers up the side of a glass "a little stream of clouds, channel up her belly from her tail to her head, which never return again the same way, but probably go backwards again from the head down the back to the tail; and thus, so long as she is in local motion they retain their circulation."[10] As soon as the snail stopped moving this circulation stopped too. Power identified the "stream of clouds" with the animal spirits, which circulated through the body. In connection with this conclusion Power set down his views on the animal spirits, which he perceived of as the most subtle kind of matter, which was "immersed or imprisoned"[11] in grosser matter throughout the universe. Within the animal body the animal spirits were freed from their prison through digestion and thereupon served their purpose in the animal "oeconomy," that is, the operation of sense and motion.

Fourth, Power deduced from his several observations, particularly the examination of crystals of salt, sand, and sugar, that the color of any object is no more than a modification of light, brought about as the light hit against the object or passed through it. Finally, he held forth on the fabric of the compound and vertebrate eye. This was meant not so much as a conclusion but rather as a reminder of the intricacies of animal anatomy and the immense terra incognita it yet constituted. He concluded his essay with the following: "Many more hints might be taken from the former Observations, to make good the Atomical Hypothesis; which I am confident will receive from the *Microscope* some further advantage and illustration, not onely as to its universal matter, Atoms; but also, as the necessary Attributes, or essential properties of them, as Motion, Figure, Magnitude, Order, and Disposition of them in several Concretes of the World."[12]

Whatever Power himself may have hoped as to the achievements of the microscope with regard to the corpuscular hypothesis, his own observations did not in any way substantiate such a hypothesis. What he did discover was that anatomical structures still existed in very much the same fashion at a level beyond the reach of the unaided human eye as they did at the macroscopic level. A definite gap exists therefore between the author's views concerning the corpuscular basis of organic matter and the smallest visible structures.

Robert Hooke

Hooke's *Micrographia* was, like Power's discourse on microscopy, essentially concerned with the fundamental structure of living bodies, that is, with the construction of living matter from inert matter. This issue is explicitly discussed, as we have seen in the preceding chapter, in relation to the first fifteen or so "Observations" in Hooke's book. In these Hooke described the appearance of artificial objects, such as a piece of cloth and glass drops, and some mineral objects as seen through the microscope, and used his findings as starting points for the development of his views on matter. In the forty subsequent observations Hooke referred occasionally to his original theme but concentrated primarily on the description and explanation of the various structures of the specimens visible with his microscope.

The first and major part of Hooke's activities in microscopy, which culminated in the publication of his *Micrographia,* was executed during the year 1663, more or less ordered by the members of the Royal Society. In the records of that body it is reported that "Mr. Hooke was solicited" on 25 March 1663 "to prosecute his microscopical observations, in order to publish them,"[13] and the members required Hooke to bring some microscopic observations each week. Before that date Hooke had already presented some microscopic observations, for example, of snowflakes and crystals of frozen urine.

The members of the Royal Society, stimulated by Leeuwenhoek's letters containing his descriptions of the recently discovered infusoria and the spermatozoa, also launched Hooke's second period of microscopic activities during the latter half of the 1670s. Having heard the contents of

Leeuwenhoek's famous letter of 9 October 1676 describing the infusoria, then usually called animalculae, Nehemiah Grew was asked on 5 April 1677 to confirm Leeuwenhoek's observations by repeating them. In the succeeding months nothing was heard from Grew on this subject.[14] Finally, six months later Hooke was requested to make a microscope like the one Leeuwenhoek used. The first time a sample of an infusion was inspected with this microscope, on 8 November 1677, the members could perceive nothing in it, but at their next meeting in the following week the animalcules were finally observed. On 6 December a compound and simple microscope were compared by examining a sample containing the animalculae, and it was found that the simple microscope performed much more efficiently.[15]

Over the next few years, several other subjects from Leeuwenhoek's letters were investigated during the meetings of the Royal Society, particularly the texture of muscle fibers and the sperm of various animals. For Hooke these activities resulted in a lecture read to the Royal Society on 14 March 1678 concerning his experiences with methods for preparing specimens, and with means of constructing single-lens microscopes; this lecture was later that same year published in his Cutlerian Lectures, entitled *Microscopium*. In this treatise Hooke dealt with the various types of microscopes (now appreciating the single microscope, which he had formerly denigrated), their most advantageous use, and microscopic techniques. He professed to be somewhat vexed that Leeuwenhoek refused to divulge his methods of research; his secretiveness had caused Hooke more trouble than had been necessary in preparing the specimens for the meetings of the Royal Society. Hooke therefore deemed it useful to describe the methods he had developed for his own use in order that anyone wanting to perform similar investigations would have the advantage of his experience.[16]

In approximately 1680, after some years of infrequent demonstrations and subsequent discussions, the subject of microscopy was dropped altogether by the members of the Royal Society. Many years later, in 1692 and 1693, Hooke read two lectures concerning the history and use of the microscope. In these lectures Hooke let it be known that he was very disappointed over the neglect of microscopic research in recent years, and recommended the microscope as "one of the most proper Ways of discovering the true Texture and Mechanism of Bodies."[17]

Hooke's *Micrographia* is composed of sixty separate observations, of which eight concern topics that have no, or hardly any, relation with microscopic science. Of the remainder five are devoted to man-made objects such as a piece of silk and the edge of a razor; five to mineral objects such as particles of sand and ice crystals; fifteen to vegetable objects, among them the famous piece of cork in which Hooke discovered the walls of what is now known as the plant cell; and twenty-seven to animal objects, mostly insects.

Micrographia has been the subject of a number of historical essays. In some it is valued as just what it seems to be: an exquisite but rather haphazard collection of separate microscopic observations on a number of unrelated objects.[18] Yet because of its exhaustive descriptions, and especially because of its superb illustrations, *Micrographia* is appreciated in these studies as a pinnacle of seventeenth-century microscopy. Others have studied the contents of *Micrographia* in relation to various contemporary issues. Historians of science appreciate it as an enquiry into the "visible substructure of life," in which the microstructure is examined as the cause of physiological processes.[19] But it can also be read as an apology for the aims and methods of the Royal Society. *Micrographia,* it is argued, was deliberately designed to document and promote the use of empirical methods and was meant to underscore the good that comes from the new science.[20]

However, there is even more to *Micrographia* than that. It also bears the mark of Hooke's views as to the proper prosecution of scientific enquiry. The choice of his specimens may have been haphazard; most of them were easily available, and some were included because the members of the Royal Society insisted on them. Yet the presentation of Hooke's observations was certainly not haphazard. On the contrary, the succession of various objects was determined by Hooke's ulterior arguments, which were hidden beneath the superficial aspects of *Micrographia,* and to which we will turn presently. The selected subjects were observed with due respect for the flaws and illusions of the optical system. The true appearance of an object, Hooke found, was "no easie matter to determine, as he that examines it shall find; for every new position of it to the light makes it perfectly seem of another form and shape, and nothing what it appear'd a little before."[21]

Through careful comparison and critical assessment of the images observed in different circumstances, however, Hooke arrived at definite conclusions about the appearances of his objects. Each individual observation in *Micrographia* is composed of a detailed description of the object, supplemented by a superb illustration and followed by Hooke's speculations on the function and operation of the various structures. The impact of the illustrations accompanying Hooke's descriptions can hardly be overestimated. They were all large in size, the illustrations of the head of the drone fly, the mite, and the flea so large that they each required a double or even quadruple page. Compared to the small and indifferent illustrations of microscopic observations published in previous decades by Odierna, Borel, and, very recently, Henry Power, Hooke's illustrations were impressive not only because of their size but also because of the smallness and delicacy of the details observed on the various objects.

An examination of the chapter on the compound eye of the drone fly illustrates Hooke's approach to, and handling of, the individual specimens. The chapter began with the description of the exterior aspect of the compound eye (see figure 12). Hooke observed that the eye is composed of numerous hemispheres (he calculated some 14,000), each perfectly smooth and distributed over the surface in a regular fashion. The fact that the hemispheres are directed toward all angles elicited from Hooke the remark that a fly is a truly circumspect animal.[22] Hooke subsequently dissected the eye to study its structure. He found it to consist of an outer, transparent layer, or cornea; a clear liquid inside the cornea; and a mucous lining within the cavity of the eye. He observed that the mucous lining is differently colored in different kinds of insects, for which reason their eyes vary in color. Hooke found these three structures in the eyes of various kinds of insects, but in some the remainder of the cavity appeared to be hollow, while in others it was filled with a "musculous" or other kind of substance.

Having recorded his observations, Hooke argued that the structures investigated must indeed be the eyes of the animal because they consist of a cornea, a transparent liquid, and a retina (i.e., the mucous lining) just like the eye of a vertebrate. Moreover, there is no other structure present on the head of the fly that might conceivably be the organ of sight. Finally, the separate hemispheres are very much like the stalked eyes of crabs and other crustacea. It is easy to prove that these stalked eyes are the organs of sight by simply cutting them off, whereupon the animal moves about blindly.

The remainder of the chapter on the fly's eye is devoted to Hooke's views on the operation of the eye and his speculations on the visual faculty of insects. On the basis of the laws of refraction Hooke concluded that each separate hemisphere refracts all the incident parallel light rays into a point on the retina, and that each hemisphere therefore is equal to a perfect eye. Consequently, "There are as many impressions on the *Retina* . . . as there are Pearls or *Hemispheres* on the cluster."[23] Yet because the protuberant surface of the hemispheres cannot refract lateral rays into a dis-

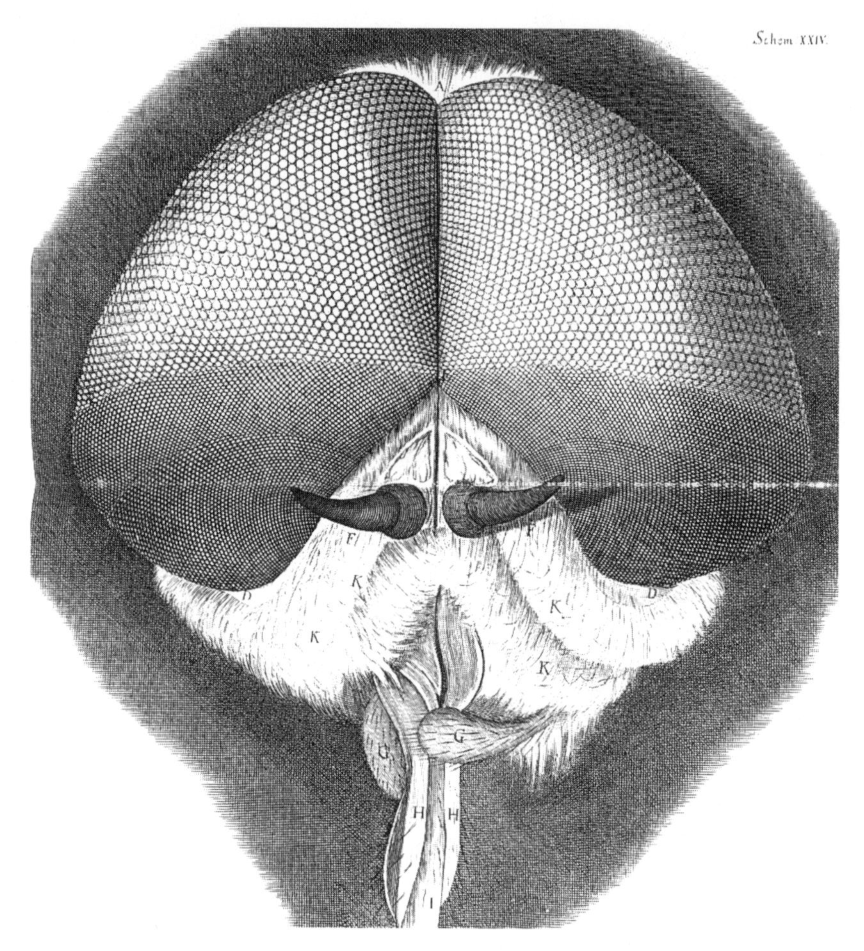

Fig. 12. Hooke's illustration of the compound eye of the drone fly, table 24 in his *Micrographia* (1665).

tinct point, a distinct image of a particular object is only formed through the hemisphere in whose optical axis the objects are situated. Taking into account that hemispheres are directed toward all angles, it would seem logical to conclude that an insect distinctly sees all the things around it at the same time. However, Hooke thought otherwise: he assumed the insect to have an "observing faculty," which it trained on the object in which it was most interested.[24] The "observing faculty" was introduced by Hooke as an analogy of the movement of the eye in man, who thereby directs the optical axis of his eyes toward the object he wants to see distinctly. An inquiry into the nature of the observing faculty, Hooke hastened to add, required another place and "much deeper speculation." It was nevertheless clear, at least to Hooke, that by the shift of the observing faculty from one "pearl" to another the insect may determine the position of a certain object, and could judge the distance to that object from the distance between the two hemispheres, one in each cluster, forming a distinct image of that object.

In a final paragraph Hooke marveled at the minuteness of the sensitive parts, located in the retina, in insect eyes. It made him wonder "how exceeding curious and subtile must the component parts of the *medium* that conveys light be,"[25] but he despaired of ever discovering the "determinate bulk" of these particles with an optical instrument.

Hooke's findings on the construction of the compound eye, as well as his conclusions about its operation, are similar to those of Odierna. It is not known whether Hooke knew the contents of Odierna's booklet concerning the eye of the fly at the time he wrote his *Micrographia,* but if he did not, he certainly came to know of its existence later, because he mentioned it in one of his lectures on the microscope given in the 1690s.[26]

Whereas Hooke always provided detailed descriptions of the outward appearance of the object, a complementary dissection was seldom undertaken. Hooke refrained from dissection because he wanted to investigate the "workings of Nature" and the microscope offered the unique opportunity to study nature

> through these delicate and pellucid teguments of the bodies of Insects acting according to her usual course and way, undisturbed, whereas, when we endeavour to pry into her secrets by breaking open the doors upon her, and dissecting and mangling creatures whil'st there is life yet within them, we find her indeed at work, but put into such disorder by the violence offer'd, as it may easily imagin'd, how differing a thing we should find, if we could, as we can with a *Microscope,* in these smaller creatures, quietly peep in at the windows, without frighting her out of her usual bays.[27]

These lines were written with much feeling because Hooke had been repelled by the suffering he had caused dogs while performing vivisection, in the course of his investigations of respiration.[28] Many years later he still treasured the microscope as a method particularly suited to wresting nature's secrets from her, without first destroying her structure. Indeed, as a method of enquiry the microscope is to be preferred to, for instance, chemistry because

> the method of Dissecting and anatomizing a body is a much more probable and has experimentally proved a more effectual way to Discover the construction and make and use of the Constituent parts of animated bodys and the uses to which they are subservient, and as living Dissection, or inspections and Experiments & observations made whilst nature is yet acting, are more informing than by Destroying the life of the animated body & beating it all to a mash in a mortar with water or any other substance & then examining of the Composition.[29]

Although the mention of the "use of the Constituent parts of animated bodys" in the above quotation might be understood to refer to the physiology of plants and animals, Hooke's speculations rarely touched on that subject. As an experimental philosopher mainly concerned with physics, he was predominantly interested in the physical properties of his organic specimens. Some of these properties could be explained on the basis of Hooke's microscopic analysis of the structure of the objects. Thus the blackness of charcoal resulted from the porosity of that substance, since the pores totally prevent reflection of the light. From the structure of cork, that is, a congeries of air-filled, tightly closed boxes, it followed that cork must float on water. The colors of a peacock's feather are caused by the small lamellae, which are closely imbricated on the side branches of the feather, and which reflect the incident light.[30]

Whenever Hooke speculated about the operation of the structures he had investigated, his conjectures concerned structures that operated mechanically, such as the stings of the stinging nettle. Hooke observed that these consist of two parts: a needle affixed to a liquid-filled sac. Any pressure against the needle caused it to be pressed into the sac, whereupon the fluid was discharged through the needle, which therefore acted as a syringe.[31]

Another mechanical action carefully studied by Hooke was the movement of the hairs on the "beard" of oats. These vegetable structures may be made to move by moistening or drying them. Hooke found these hairs to consist of a spongy and another, more solid, substance ingeniously twisted together. The spongy part attracted moisture easily, and the unique construction of the hair ensured that it shrank in dry air and stretched in moist air.[32] In the construction of his hygroscope Hooke made use of the great sensitivity of the oat beard to moisture in the air.

In *Micrographia* Hooke was not concerned with traditional physiological problems such as respiration, digestion, generation, and movement, except in the most general of terms. In connection with the diet of the bookworm, for instance, he wrote

> I cannot chuse but remember and admire the excellent contrivance of Nature, in placing in Animals such a fire, as is continually nourished and supply'd by the materials convey'd into the stomach, and *fomented* by the bellows of the lungs; and in so contriving the most admirable fabrick of Animals, as to make the very spending and wasting of that fire, to be instrumental to the procuring and collecting more materials to augment and cherish it self, which indeed seems to be the principal end of all the contrivances observable in bruit Animals.[33]

Hooke's observations were generally limited to a careful description of the exterior of the object, the section of cork and dissection of the drone fly's eye the exception rather than the rule. The measure of details resolved by Hooke's research ranged from the scales of fish to the cells in cork, and a number of the structures he discovered drew inspired guesses as to their function or operation. Such guesses often relied on, or were supported by, mechanical parallels.

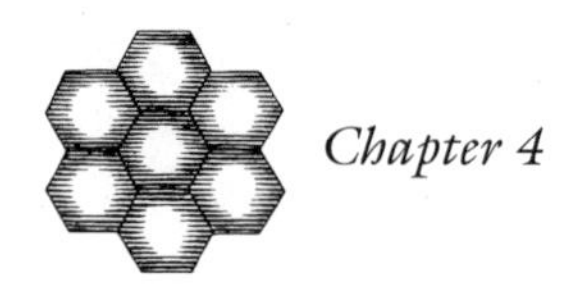

Chapter 4

The "Animal Oeconomy"

DURING THE Scientific Revolution the methods of investigation as well as the explanatory framework in physiology (or "animal oeconomy" as it was then usually called) rapidly developed. The investigation of physiological phenomena was chiefly based on traditional anatomical techniques: description and dissection, an approach commonly dubbed the "anatomical method." This method included a number of enquiry routines: dissection of individual specimens, comparative anatomy, and a survey of relevant pathological cases. The objective of the "anatomical method" was to link the particulars of its structure to the function of a specific organ. In the course of the seventeenth century investigators gradually began to make use of various new techniques, such as use of the microscope and the syringe, with which to explore the finer details of the organs of man and animals.

The interpretation of the physiological processes, on the other hand, was placed on a wholly different footing by the adoption of the mechanical mode of explanation. As a result the traditional explanation of the physiological processes on the basis of the imputed "qualities" of various organic structures was substituted for explanations based on the demonstrable characteristics of organic matter.

Crucial to the revolutionary changes in the explanatory framework was René Descartes's challenge to the traditional explanations for the operation of the human organs. He put forward his views on the physiological phenomena of man in his *Traité de l'homme,* finally published in 1664, in the form of a compelling description of the parts and processes of a mechanically operated model of the human body. Since Descartes's anatomical experience was rather slight, his notions were mainly derived from extant publications. Whereas Descartes had not worried overmuch about whether or not the mechanical explanations he presented for various organs fitted the known anatomical details of the latter, the need for sound anatomical investigations within the context of Cartesian physiology was strongly argued by Niels Stensen in his celebrated and influential lecture entitled *Discours sur l'anatomie du cerveau.*[1] Stensen heeded his own advice and executed painstaking research on the glands and muscles, which was published in his *Elementorum myologiae specimen* (1667), in which anatomical investigation and mechanical interpretation were equally matched. Judging by the number of similar studies published from the early 1660s on, Stensen had voiced a widely acknowledged concern.

The mechanical interpretation of the visible characteristics of organic matter was certainly revolutionary even if the method of investigation used—dissection and description—was essentially only a sophisticated and technologically advanced version of traditional practices. Furthermore, during this period chemical analysis of organic matter was first introduced into physiological research. On the basis of—admittedly very primitive—chemical examinations of organic substances, conclusions concerning the operation of the organs were put forward, involving the interaction between organ components. As an example of such studies Regnier de Graaf's research on the pancreas may be mentioned, which he published in his *Tractatus . . . de succi pancreati natura et usu* of 1664. The interpretation of physiological phenomena on the basis of chemical analysis was necessarily of a wholly different kind to that advanced on the basis of visual inspection. Whereas the findings of visual inspection were usually found to be compatible with a mechanical interpretation, the data of chemical analysis called for a corpuscular interpretation, in which physical phenomena are explained in terms of the mechanical interactions of the invisible and insensible particles of matter.

Within the context of the mechanical philosophy the life processes of plants began to capture the interest of scholars. Throughout the second half of the seventeenth century it was assumed that there existed a close analogy between the mechanisms underlying the physiological processes in animals and plants. As the details of the structure of various vegetable parts were discovered, largely through microscopic investigation, much effort was directed toward linking these structures with current notions of animal physiology. In this chapter we will therefore discuss both the animal and vegetable "oeconomy."

Mechanical Physiology: The Muscle Fiber

Descartes's mechanical explication of muscular action, explaining how contraction was brought about by an influx of *spiritus animalis* rushing down the nerves into the muscle, which thereupon was inflated and consequently also shortened, formed the starting point for subsequent theories. In 1664 Descartes's theory was somewhat refined by William Croone in his *De ratione motus musculorum*. Croone, like Descartes, attributed muscular action to the inflation of the muscle. During a meeting of the Royal Society in 1661 he demonstrated by pouring water into an empty bladder on which a weight was suspended that an inflated muscle could exert substantial force.[2] As the bladder filled, the weight was lifted. However, Croone disagreed with Descartes as to the cause of inflation of the muscles. He knew from experience that the nerves do not appear as hollow channels suitable for the mass transportation of animal spirits or a nervous fluid. As the swelling of the muscle could therefore not be attributed to a substantial influx of some kind of material substance, an internal force of expansion must be the cause of inflation. Appreciating the well-known fact that a continuous supply of blood is indispensable for the enduring action of the muscles, Croone proposed that minimal amounts of *spiritus animalis,* which he himself took to be a very refined kind of matter, interacted with the blood within the muscle tissue. The contact between these entities he thought to result in an ebullition or effervescence, which caused the swelling and consequently the contraction of the muscle.

The idea of an explosive reaction between blood and *spiritus animalis* or *succus nerveus* as the cause of muscle contraction was widely discussed at the time, particularly within the circles in which Croone moved. In fact, Croone's *De ratione motus musculorum* was (anonymously) published in 1664 as an annex to Thomas Willis's *De cerebri anatome,* who advanced a rather similar scheme to Croone's.

The most serious objection to be advanced against Descartes's theory was that muscles do not increase in volume during contraction; on the contrary, their volume slightly decreases. An elegant demonstration performed by Jan Swammerdam in ca. 1664 unequivocally proved this point. Swammerdam had isolated the muscle of the hind leg of a frog and left part of the innervating nerve intact. He placed this nerve-muscle preparation in a glass cylinder filled with water, and put a stopper in the top. The nerve could be irritated by means of a wire connected to it, which passed up the side of the stopper. On stimulation the muscle contracted and Swammerdam saw that the level of the water sank. He concluded "that a Muscle during contraction does not inflate or expand through the supposed influx of animal spirits and their effervescence, but that a Muscle during contraction does rather deflate, or to express my thoughts better; that it takes up less space."[3]

He also found that he could touch the severed nerve to the isolated muscle many times, and that each time it would contract again. Swammerdam's conclusion, supported by other observations, was that the nerve does not discharge any *spiritus animalis* into the muscle in order to inflate it. He wrote, "It seems to me not unreasonable to conclude that nothing but a simple and natural touching or irritation of the Nerves is necessary for the movement of the Muscles: whether this takes place in the Brains, the Spinal cord, or somewhere else."[4]

At the same time Swammerdam's friend Niels Stensen made a thorough investigation of the macroscopic structure of the muscles and the changes in their appearance during contraction. Stensen explained methodically how each muscle is composed from innumerable motor fibers as well as blood vessels and nerves. These fibers are geometrically arranged so as to form a parallelepiped (the body of the muscle) bordered by two tetragonal prisms (the tendons). Applying mathematical laws, Stensen deduced that during contraction the motor fibers are rearranged so as to bring the tendons closer together, a mechanism that caused the width of the muscle to increase, while its volume remained constant. Stensen's theory therefore agreed with Swammerdam's observations.

On becoming aware, through Stensen, of Swammerdam's experiments, Croone began to rethink his original scheme since he would have to explain the contraction of the muscle without concomitant swelling. He solved that problem by assuming that the muscle fibers were formed from numerous small bladders arranged one after the other like pearls on a string. He developed this idea in a course of lectures presented to the Royal College of Surgeons in 1674 and 1675. Eventually these lectures were published in the 1681 issue of Hooke's *Philosophical Collections.* Croone felt himself fully supported by Leeuwenhoek's research, as the latter had recorded that he had seen that the texture of the fibers of the flesh consisted of innumerable microscopically small vesicles or globules.[5] Leeuwenhoek had studied the microscopic structure of muscles beginning in 1674. At that time he described the construction of the muscle fibers as a row of globules, secured within a very thin membrane.[6]

Robert Hooke heartily disagreed with Croone on the microscopic structure of the muscle. During a meeting of the Royal Society during which muscular action was discussed, he said he had observed that the fleshy part of muscle consists of an infinite number of exceedingly small round pipes. However, he agreed with Croone insofar as he also thought it possible that contraction resulted from the swelling of these pipes. At the same meeting Grew declared that he had observed "that the fleshy part of a muscle was divided into a sort of long parallelepipeds by the cross interweaving of small membranes and vessels crossing the said fleshy part."[7]

Croone immediately identified the fleshy parallelepipeds with his chain of bladders. Although the remaining members present desired some more definite "ocular demonstration" from each of the disputants, the subject was dropped altogether for some years until Hooke returned to it in 1678. Then he showed the members of the Royal Society during a meeting a microscopic experiment that revealed that the muscle in the claw of a lobster consists of numerous very fine filaments, each of which appeared to the spectators in the shape and figure of a wreathed pillar.[8] A few weeks later Hooke performed an experiment meant to support his contention that a filament with such a construction might indeed exert force. He knotted a helical gut and a piece of string together at both ends. As air was blown into the gut, the string was shortened because the expanding gut wound itself around the string, causing it to curve around its surface. Conversely, as air was released from the gut, the string lengthened again as the gut returned to its original shape.[9] After the demonstration Hooke promised the members that he would perform an experiment to support the "string of bladders" hypothesis of Croone. Having done so the very next week, the members discoursed on the cause of the swelling of the bladders, some advocating the idea of air expanding or contracting as it grew hot or cold as a result of nearby effervescing liquids, others dispensing with air as an intermediary agent for the fermenting heterogenous fluids to cause the expansion of the fibers.[10]

From these experiments it appears that Hooke delighted in devising models inspired by the microscopic appearance of muscle fibers with which to demonstrate that the imagined construction could indeed be the mechanical cause of muscular action. In his *Microscopium,* however, he warned (as a side issue to his description of the muscle as a bundle of innumerable small fibers) that it is not really possible to make a direct comparison between gross, macroscopic processes with the operation of refined microscopic structures.[11] He alluded to the fact that such a bundle is very much stronger than a single strand of material of equal width.

Independently of Croone, Giovanni Alphonso Borelli had arrived at a similar model of the ultimate structure of the muscle on the basis of a rigorous mathematical analysis of the muscle and its performance. The muscles are composed from threads, he wrote, but the action of the muscles can only be explained if it is supposed that these threads are arranged in such a way that the "fibers of the muscles are a series of little mechanisms, porous or rhomboidal in form, like a chain composed of threads in a rhomb."[12]

He argued that, on contraction, each little rhomb changed its form in such a way that its transverse axis became larger while the horizontal axis became smaller. Thus the diameter of the muscle became larger while its length shortened. This theory was advanced in the first volume of Borelli's *De motu animalium*, which was prepared quite some time before the second part,[13] although both parts were printed in consecutive years. In the second part Borelli discussed his ideas concerning the cause of contraction, in which he followed Croone in most respects. Borelli had also studied the fibers of the muscle through the microscope, and observed that the interior of the fibers was filled with a spongy material.[14] Even though this observation did not actually substantiate his theory of muscular action, it did not contradict his views.

Croone may have felt supported by Leeuwenhoek's preliminary observations on the muscles, but Leeuwenhoek exploded this proof once he had revised his description of the texture of muscles in 1682.[15] By that time he had realized that what he had formerly taken for globules were in fact often cross sections through fibers. Leeuwenhoek's subsequent research into the precise structure of the muscle was performed on samples from a large number of animals: oxen, whales, fish, mice, crawfish, shrimps, fleas, flies, and so on. He found them all to be of the same basic structure, that is to say, the macroscopic fibers each consisted of a great number of smaller fibers, or filaments. Each of the filaments he observed to be surrounded by rings or "wrinkled contractions." Prompted by a query from Hooke, writing in his capacity as secretary of the Royal Society, Leeuwenhoek specifically studied the muscles of a shrimp (for lack of a lobster) so as to confirm or invalidate Hooke's observation that the lobster muscle appears as a "string of pearls" through the microscope.[16] He did not find any difference between the shrimp muscle and the general scheme, and he thought that the "string of pearls" effect had been brought about by the light falling obliquely on the "wrinkles" of the muscle fibers. In the last decades of his career Leeuwenhoek perfected his earlier description of the "wrinkles," which he had formerly thought to form circles on the surface of the fibers. On closer inspection he thought they were actually arranged spirally.[17] Leeuwenhoek observed that the membranes around the muscle fibers not only formed one continuous whole, but that they continued into the tendons, thus uniting the muscle fibers with the filaments of the tendon.[18]

Leeuwenhoek's interpretation of his microscopic view of the muscle as regards muscle action is rather curious. He thought that the active muscle was in fact longer than the relaxed muscle. He phrased it thus: "When we see the circular wrinkles in the muscle fibers we must conclude that the muscles or rather the muscle fibers are at rest; and when the muscle moves, or rather extends, then the circular wrinkles disappear."[19]

This passage means that Leeuwenhoek thought of muscle action as a process causing the elongation of the fibers rather than their contraction. Leeuwenhoek's theory is reminiscent of Claude Perrault's theory of muscle action. The latter held that the membranes surrounding the muscle fibers are elastic, and that they keep the fibers in a shrunken state while the muscle is inactive. As soon as an impulse from the nerves acts on the muscles, this network of membranes relaxes so that the fibers can elongate. In the case of a limb movement this means that as the antagonist relaxes and consequently becomes longer, the protagonist, as a result of the elasticity of the membranes, necessarily shrinks and brings about the movement. Perrault published this theory in 1680 in his *Traité de physique*. It is therefore certain that Perrault's ideas were not stimulated by Leeuwenhoek's microscopic observations. Conversely it is very doubtful if Leeuwenhoek knew about Perrault's ideas.

The origin of Leeuwenhoek's notions on muscle contraction may perhaps be traced back to his poorly digested knowledge of Cartesian physiology. Descartes held that the heart expanded as the inflowing blood was heated as a result of the heat of the heart. He, like Leeuwenhoek many years later,[20] regarded the expansion of the heart as the active part in cardiological movement. On the other hand it has been suggested that Leeuwenhoek's ideas concerning muscular action issued from his observations on the sessile infusorium *Vorticella*, which characteristically may suddenly retract its body by contracting its stalk.[21] The contracted stalk presents a striated appearance, which disappears as the organism shoots up again.

However rapturous posterity may be about the excellent quality of Leeuwenhoek's observations on the texture of muscles, his contemporaries were often unable to verify his findings; even Hooke could not substantiate the observations he communicated to the Royal Society during the 1680s.[22] Neither did Leeuwenhoek's reports exert much influence on contemporary and subsequent writers, perhaps because his data, or rather his interpretation of these, were not suggestive, unlike the "string of pearls" construction, of a mechanical mode of operation.

The way in which the details of Leeuwenhoek's results failed to be incorporated in contemporary knowledge is illustrated by Weyer Willem Muys's microscopic research on the muscle. Muys elaborated on Leeuwenhoek's data to a certain extent, and among other things he attempted an exact measurement of the various composite muscle fibers of different sizes.[23] He distinguished between no less than eight different compound fibers, each with its specific diameter. The smallest of these he identified as the basic structural and functional element. Muys described and represented this elemental fiber as a string of pearls.[24] From his book it is clear that Muys was fully conversant with Leeuwenhoek's observations and views, and therefore no doubt aware of the fact that Leeuwenhoek's description of the elemental fiber differed significantly from his. He nevertheless adopted a basic element that was in keeping with the prevalent theory of muscular action, founded on the string-of-pearls construction of the fibers. However, Muys did not adopt this theory on theoretical grounds. He had in fact observed, to his own satisfaction at least, that the elemental muscle fiber is a hollow tube.[25] He had injected the small arteries of a frequently rinsed muscle with a colored fluid, and observed microscopically that even the smallest muscle fibers, the tubuli, were filled and tinged with that fluid, while the interstices between the fibers remained blank. He therefore concluded that the smallest fiber or tubulus is hollow, and that the extremities of the capillary arteries open into them.

An equally enigmatic connection between observation and theory is apparent in Baglivi's theory of muscular action. Giorgio Baglivi, a student of Malpighi's, also studied the structure of the muscles microscopically and, like Leeuwenhoek, observed that both the muscle fibers and the surrounding membranes were formed from filaments. Having subjected the muscle samples to various preparatory techniques, including treatment with alcohol and corrosive fluids, he established, that the filaments in the membranes surrounding the muscle fibers run in all directions through one another, while in the muscle they run parallel.[26] Baglivi imagined that, as the muscle contracted, the fibers glided over the particles of the blood, which acted as rollers. He therefore assumed that the particles of blood circulated freely between the fibers of the muscle. In view of Baglivi's methods of preparing his specimens, this characteristic detail of his concept was definitely not founded on observation, but the extent of his microscopic observations did not contradict such a mechanism.

Marcello Malpighi

Between 1661 and 1666 Malpighi published a number of short treatises on the anatomy and function of several of the organs of the human and animal body. Some of these were issued separately and others collected into small volumes. Successively the lung, tongue, skin, and brain were dealt with and, separately, the cerebral cortex, liver, kidney, and omentum.[27]

Malpighi first directed his attention toward the lung. He found this organ, rather than the reputedly indifferent fleshy mass, to have a complex structure, consisting entirely of thin membranes shaped into vesicles that sprouted from the finest branches of the trachea. Of particular significance for the development of Malpighi's ideas concerning the mechanism of physiological processes was his discovery that the blood vessels in those membranes branched out into numerous minute vessels. These vessels formed a reticular pattern on the walls of the vesicles, and as the ramifications of the artery branches passed into the prolongations of the veins, this network constituted in fact a connection between the arteries and veins of the lung. In Malpighi's view the operation of the lung involved both the extensive ramifications of the blood vessels and the air-filled vesicles. He surmised that the various constituents of the chyle and the blood became thoroughly mixed in the lung as a result of the distribution of the blood into narrow passage ways, supported by the perpetual expansion and contraction of the vesicles as the animal breathes.

In his subsequent investigation of the texture of the liver, kidney, and spleen Malpighi discovered numerous round bodies in each of these organs (in modern terminology these bodies are the lobules, glomeruli, and splenic lymph follicles, respectively) in addition to a series of other novel details concerning their main structure. These spherical bodies, he observed, were enveloped in minute blood vessels and, in the cases of the liver and kidney, branches of the bile duct and renal tubuli, respectively. Malpighi postulated that these bodies were glands and that they in fact constituted the functional units of each organ. He conceived of the glandular bodies as hollow follicles or acini, supplied with an excretory duct. With reference to the lung vesicles, in which it is observed that "the radicles of the veins take their origin from the same place in which the terminal arteries end,"[28] Malpighi concluded that the glandular follicles were intimately connected with both arteries and veins. The nature of that connection was instrumental in separating parts (such as bile or urine) from the blood. Indeed, the common work of the glandular follicle and the blood vessels consisted of separating the glandular product from the blood. Although Malpighi tried, in the case of the kidney, to confirm experimentally that a connection exists between the globular bodies, the arteries, and the veins, and that these bodies opened into the renal tubuli, he succeeded only insofar as he demonstrated that a colored fluid injected through a branch of the arteries colored the globular bodies as well.

Malpighi published two treatises on the brain: *De cerebro* of 1665 and *De cerebri cortice* of 1666. In the first of these studies he investigated primarily the white matter of the brain, which he found to consist of numerous fine filaments. These were readily observed, particularly in the brain of fishes, where the structure was reminiscent of a "white ivory comb." "I see," Malpighi wrote, "clearly that this whole part of the brain is divided in flattened, round fibers."[29] At that time he was not sure whether or not these filaments were hollow, but he thought it likely since the cortex was supposed to produce the nervous liquid, which was discharged into the filaments. In his second treatise Malpighi tackled the question of the structure of the cortex, and demonstrated that it is composed of innumerable glandular bodies. With the aid of the microscope Malpighi observed that, when ink was poured onto the surface of the brain, numerous blank areas showed up against an ink-stained network, and he concluded that the blank areas represented glandular bodies. Malpighi's pertinent observations as to the fabric of the brain therefore consisted of the establishment of the presence of fibers and globular structures, the latter enveloped by the ultimate branches of the blood vessels. From these data Malpighi concluded that "the cortex is a crop or congeries of very small glands. . . . The inner part of each gives off a white nerve fiber as its excretory vessel and the white medullary substance of the brain is formed by the gathering of many such fibers into bundles."[30]

Malpighi's studies on the brain were supplemented with his investigations concerning the structure of the tongue and skin, in both of which he discerned a horny and a mucous layer, covering papillae underneath. Malpighi found the papillae to be extensions of the nerves, and concluded that taste and touch are experienced when the papillae are stimulated by their appropriate agents. In sum,

> It is certain that the brain is composed of hollow cords which receive fluid that makes them more or less tense, and since these cords are arranged like the strings of a lute, it follows when a slight movement occurs in the outer sense organs (which are the ends of nerves) as a result of agitation set up, for example, by light in the eye, by air in the ear, by salt on the tongue, by solids on the skin, and by internal fluids at the root of nerves, a tremor necessarily occurs in the nerves themselves and then at the other end, where they are arranged and stretched, and this will be the physical motion of the internal senses, the properties of which can in all probability be discussed by analogy with mechanics.[31]

Descartes had voiced his views concerning nervous action in much the same terms in his various writings concerning the physiology of man, notably in his *Traité de l'homme* of 1664. However, whereas it was clear to any anatomist that several of Descartes's schemes for the mechanical operation of organs were not founded on a sound anatomical basis, Malpighi's researches put such mechanical interpretations on a much surer footing. With his microscope he resolved some of the minuter structures involved in these processes, these structures appearing to be compatible with a mechanical explanation.

After the completion of *De viscerum structura* Malpighi devoted his attention for many years to other subjects. It was not until twenty years later that he gathered the notes he had recorded over the preceding years, concerning the texture of the animal and human organs, into a conclusive treatise entitled *De structura glandularum,* published in 1689. In this treatise Malpighi claimed that "Nature proceeds by a single, simple method: she attaches to an excretory vessel one or sometimes more membranous follicles or acini, and by means of them separates a peculiar humor from the vessel and when it has been collected, expels it."[32]

A simple gland, like the miliary glands of the skin, thus consisted of but one acinus, while glands with a complex structure, like the kidney, were composed of numerous separate gland follicles. In *De structura glandularum* Malpighi surveyed all the structures he regarded as glandular, including the brain, stomach, and testes. On the basis of his anatomical research, aided by specialized techniques and the microscope, Malpighi thus came to regard the texture of the separate organs as individual variations on a universal scheme. Indeed, he judged that a great many parts of the animal body were compiled from a multitude of fine fibers, miliary glands, and extremely small vessels.[33]

When Henry Oldenburg made contact with Malpighi in 1667 by sending a letter challenging Malpighi to participate in the scientific program of the Royal Society, he explicitly mentioned the society's interest in the silk moth.[34] In response Malpighi produced a detailed study of the anatomy and development of that insect, entitled *De bombyce,* which was published in 1669. This treatise was the first of Malpighi's publications to be abundantly illustrated, each of the plates depicting at that time virtually unknown structures of the interior of the insect body. Malpighi had approached his subject with zest. He bred large numbers of these insects in his home in order to be able to observe the slightest changes during their development from the larval stage through chrysalis into moth. He dissected them during each of these stages and studied their reactions to several physiological experiments. Among the anatomical features his exposure of the tracheal system (see figure 13) and particularly the fortified structure of the tracheal tubes and the anatomy of the female generative

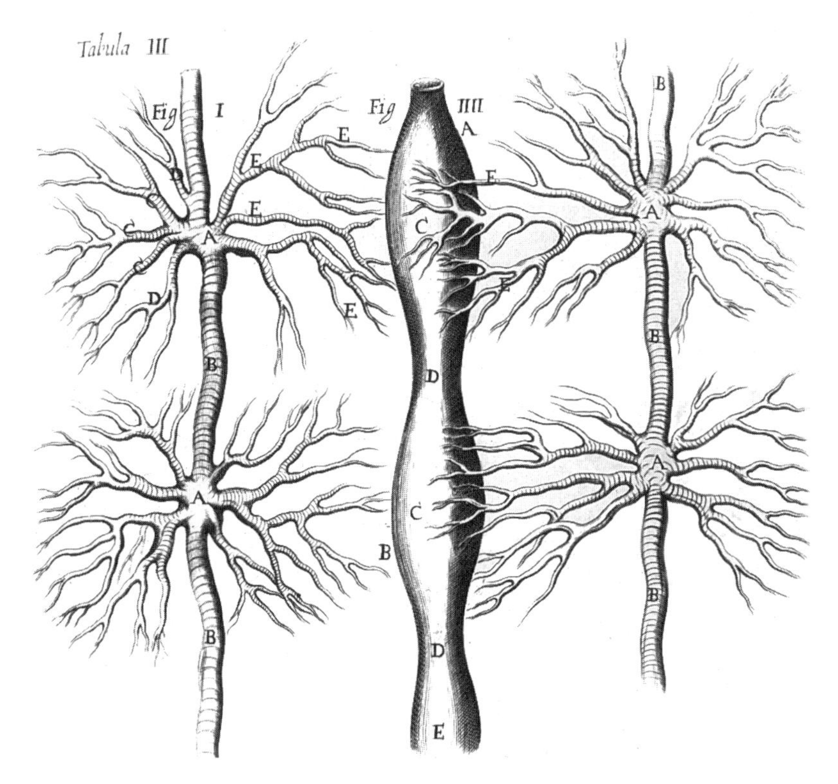

Fig. 13. Malpighi's representation of the trachea of the silk moth, table 3 in his *De bombyce* (1669).

organs of the moth are especially impressive. Malpighi proved experimentally, by sealing off the spiracles, that the tracheal system serves as the organ of respiration in insects.

Closely observing the development of the silk moth in its several stages, Malpighi noticed that the "parts belonging to, and destined to be used by the moth" are visible in preceding stages. It seemed to him that these parts gradually swelled and became more evident; they emerged as it were from the "mucous juice" of which the interior of the larva seemed to consist.[35] Malpighi was therefore confronted in *De bombyce* with the origin of the parts of the developing insects.

Subsequently Malpighi studied the development of vertebrates, choosing a traditional but convenient object: the chick embryo. This research was to some extent motivated by the wish to elucidate the component parts of living matter, which might in principle be more easily accomplished during development than in a full-grown organism. For, reasoned Malpighi, "in building machines the artisan is wont to fashion the several parts by means of apparatus previously at hand for the work, so that the things which must afterwards be fitted into the assemblage may first be viewed as separate elements."[36]

Despite the numerous rewards of Malpighi's subsequent embryological research this particular premise was not confirmed. Studying the early development of the embryo, Malpighi described numerous details such as the neural fold, somites, optical vesicles, cardiac tube, and aortic arches. He studied the appearance of the heart with special care because that bright red pulsating speck was clearly distinguishable to the naked eye before any other structure became visible, and was moreover traditionally equated with life. Malpighi, however, found that the rudiments of the heart were visible (through the microscope, that is) before the liquid propelled through it turned red. In fact it was the very movement of those parts by which he detected them. He concluded therefore that "the structure of the heart is made evident solely by its motion, and that, although weak . . . the heart existed before this time in a state of rest."[37]

One of Malpighi's most misused results, in view of the subsequent discussions concerning the issue of preformation, was no doubt the observation that the first beginnings of the embryo were visible in unincubated eggs. Even though this observation was later often cited in support of preformation, it was conveniently overlooked that Malpighi also observed that nothing of the sort could be distinguished in unfertilized eggs.

He therefore wrote, cautiously, that one could surmise that the chick lies concealed in the egg, but he himself had not then come to a definite conclusion concerning this issue because "so intricate and hidden are the handiworks of Nature, that, investigated with the help of the senses though they may be, they can nevertheless easily deceive us (me at least) because they are concerned with the very smallest objects."[38]

Malpighi's interest in the developmental process was again manifest in his investigations of the formation and germination of the seed and the growth of the seedling, which formed an essential part of his plant research. Among a host of detailed observations and some speculations on these topics, Malpighi described the plant embryo contained in the seed and recognized that it consisted of two leaves (he did not discern the fundamental difference between di- and monocotyledons), a stalk, a bud, and a root. The seed, therefore, is to be regarded as a compendium of the future plant, a plant reduced to its principal parts. He also observed in the buds of mature plants the undeveloped leaves of the future shoots. Malpighi dwelt at some length, at different occasions, on the fundamental similarity between the seeds, the buds, and the saccules or vesicles observed in animal development. Just as the seed and the bud contained the rudiments of specific parts of the future individual, so did the saccules (imaginal discs) observed in the pupa of the silk moth. Malpighi described these saccules as "the peculiar cavities which contain the primordia of each of the parts."[39] Much the same could also be said for the budlike structures such as the optical vesicles, seen in the early chicken embryo. On the basis of these considerations Malpighi concluded that the developmental process invariably started from a "compendium," contained within a baglike structure, which presently incorporated additional matter, gradually grew in size and eventually became visible. Thus Malpighi reduced to the same denominator his multifarious observations on the development of the chicken, of the silk moth, and of a range of plants, emphasizing the analogies rather than the differences. Nature, he wrote,

> brings the structure of *perfect animals,* which emerges from the egg, through continuous increase and nutrition to its proper size, crowding new particles to earlier and already preexisting particles, but in such a way that in whatever state of increase the same primeval form and nature are maintained, no part arising anew in animals except teeth and horns. But in *insects,* besides increase, as I said several times, parts arise, whose rough delineations were hidden in earlier stages, such as wings,

> feelers, and the like. In *plants* Nature, through the same generosity, attains not only daily growth by investing the trunk and branches with a woody wrapping, but each year new shoots emerge from young branches, whose anticipated beginning, as it were a peculiar fetus, is called the *bud* or *eye*.[40]

Malpighi concluded from his various observations on the development of organisms that living matter had acquired the first outlines of its texture before any indication of structure could be detected visually, either with the naked eye or with the microscope. Even so he did not consider the preexistence of structure in the egg or seed to be a serious possibility. In his opinion the matter in the egg and seed was structured at the time of fertilization.[41]

Malpighi's plant research constitutes the bulkiest part of his published investigations, as regards both the length of the text and the number of illustrations. Before venturing on the preparation of his observations into a publishable manuscript, Malpighi condensed the essentials of the insights he had gained on plant structure, as well as on the function of the several parts, into an essay entitled *Anatomes plantarum idea,* dated 1 November 1671. He sent this essay to the Royal Society which received it, by some caprice of fate, on the very day that Nehemiah Grew presented a copy of his recently published *The anatomy of vegetables begun* (1671) to the assembled members of the society.[42] The members were highly pleased with these endeavors and urged both men to continue their research. For Malpighi's part the encouragement of the Royal Society resulted in the dispatch of the planned manuscript in two separate batches, which were eventually published by the Royal Society as *Anatomes plantarum pars prima* and *pars altera* in 1675 and 1679, respectively.

In the first part of *Anatomes plantarum* Malpighi described the anatomy of the principal parts of the plant, successively the stem or trunk, leaf, flower, seed, and fruit, and devoted separate chapters to the bud and the formation of the seed. In the second part he described the germination of the seed and the early growth of the seedling, and then went on with the description of growths on the exterior of the plant, that is, excrescences such as galls, cancers, and saprophytes, but also natural extensions such as hairs, spines, and tendrils. The final chapter of this volume was devoted to a description and discussion of the root.

Just as in his animal research, Malpighi studied the several parts of the plant in a wide range of species. In doing so he discerned that the structure of plants is a well-nigh infinite variation on the basis of two units, that is, the boxlike utricle and the tubelike fiber. It must be emphasized that Malpighi's plant research was to a rather large extent macroscopic (or effected with low magnification), but the finer details of the plant's anatomy were investigated with a microscope. Among these details the thickenings on the walls of the spiral vessels may be mentioned, as well as the tyloses, and the layer of palisade cells of the inner seed capsule.

Anatomes plantarum may be a storehouse of novel data on the morphology and anatomy of plants, but since Malpighi's interests transcended the structural level, vegetation was thoroughly discussed as well. Drawing on his own premise that in animal nutrition the final blending of the nutrients with the blood takes place in the capillary network of the lung, Malpighi assigned a similar function to the network of ligneous vessels that extends throughout the plant. Accordingly, the fluids extracted from the soil with the necessary ingredients dissolved in them ascended through the vessels and became thoroughly mixed on their way up through these narrow pipes. Malpighi envisaged that this fluid, as it drains into surrounding utricles, is mixed with the "old sap" within those utricles, and presently through a fermentative process becomes an enriched, more refined nutriment. In his view this process was operative on several levels. Ordinary nutriment was thus formed in the stem or trunk of the plant, and also for the formation of the shoot buds. In the petals of the flowers the nutritive juice was once more subjected to the same process, to purify the sap before it was used for the formation of the plant embryo.

Malpighi had definite ideas about the modus operandi of the transportation of plant juices. In his opinion the heat of the sun, striking the soil, drives the water in the soil into the vessels in the root of the plant. The vessels in the bark and ligneous parts are so constructed as to facilitate the transportation of water upwards through these vessels. For, "The portions that join the fragments of fiber together project slightly inward and thereby play the role of valves, so that even the smallest drop is forced to rise to the highest extremity of the plant, as though by a rope or staircase."[43]

Yet it is the heat of the sun that really sets the transportation in motion. According to Malpighi the air surrounding the stem of the plant exercised greater pressure on the cortex by day because of the greater heat and thus assisted the upward movement of the liquids in the vessels.[44] The motion of the fluid was not only encouraged by the air surrounding the plant, but also by means of the air contained in the air vessels of the plant. During the heat of the day, Malpighi thought, these vessels expanded as the air within them expands, thus pressing on the sap-filled vessels, which caused the sap to be squeezed partly upwards and partly into neighboring vessels or utricles. Later in the day the emptied vessels were readily filled with fresh liquor as the air cooled and consequently contracted, thus exerting less pressure on the vessels. Malpighi therefore envisaged a mechanism of transportation that depended on both the structure of the ligneous vessels and on the action of the air surrounding and contained within the plant, a purely mechanical form of transportation.

Malpighi attributed analogical functions to certain structures in plants and in animals. Speculating on the role of the plant's principal parts in vegetation, Malpighi advanced some broad similes with animal physiology. For instance, he assigned a function to the leaves comparable to that of the skin, that is, getting rid of waste products by means of transpiration. The supposed analogy between animal and vegetable structures was emphasized by the application of the vocabulary of animal anatomy to vegetable structures. A case in point are the terms Malpighi employed for the description of the seed and the structures involved in its production, such as *chorion, amnion,* and *umbilicus.* Malpighi, therefore, used animal anatomy and physiology as a model with which to interpret plant structure. In doing so he contradicted the premise that, he said, motivated his research on plants, namely, that by studying the simple mechanisms of plants the complex mechanisms of animals might be better understood.

From an analysis of Malpighi's work it seems clear that he investigated the minutiae of organic matter in search of an explanation for the origin of the physiological processes. From an early date in his career it was a settled matter, as far as Malpighi was concerned, that these processes were the result of the mechanical actions of the finer components of organic structure. In his search for these components Malpighi, armed with his microscope, explored such virtually virgin and diverse areas as the texture of the various organs, the early stages of embryological development of the chick, the anatomy of insects, and the finer structure of plants. These investigations abundantly confirmed the structural subtlety of organic matter.

Even though Malpighi felt initially satisfied in some cases that a purely mechanical explanation for the operation of organs might be based on these findings, his data eventually pointed to an ulterior level of organization, a level that he could not reach with the means at his disposal. Although Malpighi indicated that the limiting mean in his view was the microscope, it seems to me that in fact it was rather the lack of microscopic technique that constituted the limiting mean. However, even if Malpighi felt frustrated at times by the limitations of his instruments, he nonetheless pursued his researches throughout his life. Indeed, his extant manuscripts contain numerous entries dating from the second half of the 1670s and the 1680s,[45] although he only published his *De structura glandularum* at the end of that period. These researches were largely concerned with plants and insects, and some of the observations on plants are inserted in his *Opera posthuma*, among them a detailed study of the development of the seedlings of the castor oil plant and the date palm.[46] Yet because of ill health, and perhaps the strain caused by the publications of Sbaraglia *cum suis*, Malpighi did not gather these investigations into further publications. Besides, in the document (written toward the end of his life) in which he defended his life's work against the attack from Sbaraglia, Malpighi once again emphasized the usefulness of microscopy with respect to the advancement of medicine.

In conclusion, it is plain that Malpighi started his microscopic researches with the intention of resolving the machinery of the physiological processes, a machinery that he conceived of in mechanical terms. The content of Malpighi's scientific career was therefore dominated by the mechanical philosophy. As the rather crude mechanical models of which Malpighi spoke in connection with physiological processes, such as sieves, were not in actual fact observed, the site of mechanical operation appeared to be located at a level beyond that which could be attained with contemporary visual aids. Even so, Malpighi did not doubt the premise from which he had started.

Nehemiah Grew

Grew's microscopic investigations formed the main part of the second general means in his scheme, that is, the anatomy of the plant. These studies were published in *The anatomy of roots* in 1673 (as a sequel to his *Idea*) followed in 1675 by *The Comparative Anatomy of Trunks.* In 1678 *Experiments of Luctation* was published, which was a first sample of his chemical experiments, or third and fourth general means, to appear in print. In the years 1673 to 1677 Grew presented his results concerning the anatomy of leaves, flowers, fruits, and seeds, the lectures later included in his *Anatomy of plants.* Also included were several further lectures on the chemical composition of plants and on the chemical origin of colors and tastes in plants.

To be able to study the structure of plants with the microscope, wrote Grew, "Three things are hereunto necessary; *viz.* a good *Eye,* a clear *Light* and a *Rasor,* or very keen *Knife,* wherewith to cut them with a smooth surface, and so, as not to Dislocate the *Parts.*"[47] As Grew's plant materials were composed of a relatively firm substance he did not develop any microscopic preparation techniques beyond cutting sections through various planes. By comparing the sections of various parts of the plant with those of several kinds of plants, Grew was led to the conclusion that "the essential Constitutions of the said parts are in all plants the same."[48] He distinguished two essentially different kinds of tissue: the parenchymatous and the ligneous. He had already discovered this much with the naked eye. On inspection with the microscope he concluded that the ligneous part was "nothing else but a *Cluster* of innumerable and most extraordinary small *Vessels* or *Concave Fibers,*"[49] fibers that he found to be continuous and extended along the axis of the root and trunk. The parenchymatous part, on the other hand, he found to be "a mass of bubbles" that were "all alike discontinuous."[50] Together the bubbles, or "bladders," as he more often called them, formed "*a most curious and exquisitely fine wrought Sponge.* This much the Eye and Reason may discover." He was well aware that these "bladders" formed separate entities: "The *Microscope* confirms the truth hereof, and more precisely shews, That these *Pores* are all, in a manner, Spherical, in most *Plants;* and this *Part,* an infinite Mass of little *Cells* or *Bladders.* The sides of none of them are Visibly pervious from one into another; but each is bounded within itself."[51]

Grew found the ligneous part to consist mainly of air vessels and sap vessels. The latter might contain a lympha, milk, gum, balsam, or other kind of fluid. Although he distinguished a great number of different "succiferous" (which he often called "lymphaeducts") or ligneous vessels he supposed that there might be even more kinds. Grew equated the plant vessels with the viscera in animals: as there is such a variety of viscera, there must be a similar variety in plant vessels.[52] In Grew's view, therefore, animal and plant physiology are very similar processes, requiring similar organs to be realized. Whether there was indeed such a strong analogy between plants and animals was currently widely debated, but by no means clear.[53]

All the parts of the plant are formed from the two basic elements, the "bladders" and the vessels, but in an infinitely varied way as to number, size, and configuration. In Grew's view most vessels are distinct from the parenchymatous "bladders," but some types of vessels appear to be fashioned as the result of the disintegration of the dividing walls between a row of perpendicularly stacked "bladders."[54] Grew was much impressed by the fabric-like or lacelike structure of plant material when viewed through the microscope. Based on that resemblance he developed a model for the structure of plant material (see figure 14). Over and over again he refers in his writings to the "Warp and Woof" of plant texture. He thought all plant structure to be built from finely intertwining fibers: "The whole *Substance*, or all the *Parts* of a *Plant*, so far as *Organical*, they also consist of *Fibres*. Of all which *Fibres* those of the *Lymphaeducts*, run only by the *Length* of the *Plant:* those of the *Pith*, *Insertions*, and *Parenchyma* of the

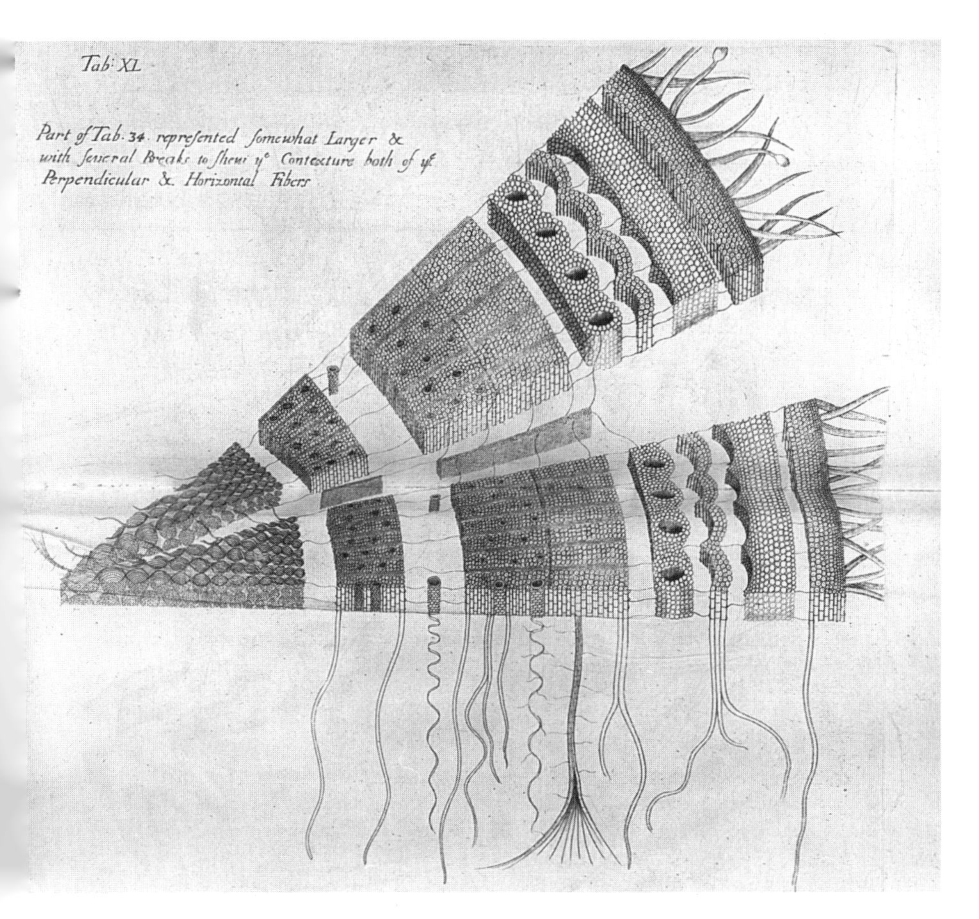

Fig. 14. Grew's representation of the construction of vegetable matter, table 40 in his *The anatomy of plants* (1682).

Barque, run by the *breadth* or horizontally: those of the *Aer-Vessels,* fetch their Circuit by the *Breadth,* and continue it by the *Length.*"[55]

These conclusions were, he thought, based on observation, but Grew was so taken up with his "Warp and Woof" model that he did not hesitate to extend the model to what he could not observe and wrote, "Even all those *Parts* of a *Plant,* which are neither formed into visible *Tubes,* nor into *Bladders,* are yet made up of *Fibers.*"[56] He postulated that "these *Fibers* themselves are Tubulous, or so many more *Vessels,* is most probable, There only wanteth a greater perfection of *Microscopes* to determine."[57] The walls of the "bladders" he explained, "are not meer *Paper-Skins,* or rude *Membranes;* but so many several Ranks or Piles of exceeding small *Fibrous Threds;* lying, for the most part, evenly over one another, from the bottom to the top of every *Bladder;* and running cross, as the *Threds* in the Weavers *Warp,* from one *Bladder* to another. Which is to say, That the *Pith* is nothing else but a *Rete mirable,* or an Infinite Number of *Fibres* exquisitely small, and admirably Complicated together."[58]

He found the wall of the air vessels (after Malpighi in his *Idea plantarum,* which Grew read in manuscript, had drawn his attention to the fact) to be spirally strengthened by "Two or More round and true *Fibres,*" fibers that were knitted together with fibers of smaller dimensions. At this point Grews's enthusiasm really ran away with him, causing him to contemplate that the several fibers are not interwoven "just as in a *Web,* but by a Kind of Stitch, as the several *Plates* or *Bredths* of a Floor-Mat."[59]

Grew wrote at length on the life processes of the plant. True to the principles of scientific investigation his intention was "as far as Inspection, and consequent Reason, may conduct, to enquire into the visible *Constitutions,* and *Uses* of their several *Parts.*"[60] He therefore aspired to an explanation of the plant's physiology "grounded upon" his anatomical findings. The chief subjects he touched on were the distribution of nutriments through the plant body, the elaboration of the "principles" (i.e., the nutrients) imbibed along with water from the soil, and the growth of the plant.

Grew conceived of a continuous circulation of fluids through all parts of the plant, which carried along the necessary nutrients (including air). This circulation had its source in the roots of the plant where water, loaded with particles from the soil, was absorbed into the plant. A continuous flow of these fluids was ensured, in Grew's opinion, by the leaves, which acted as cisterns. The water absorbed was first "strained a hundred times over" as it passed from the cortex through the "bladders" of the root into

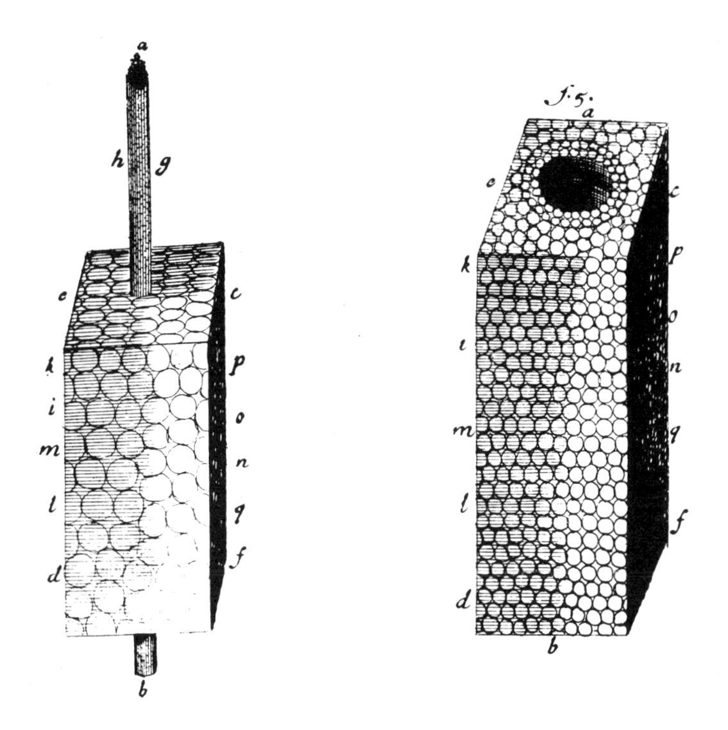

Fig. 15. Grew’s notion of the mechanism of sap transportation through the vessels of plants, figs. 4 and 5 in table 39 in his *The anatomy of plants* (1682). Both figures represent a vessel surrounded by “bladders,” which swell as they absorb water from the nearby vessel and subsequently press it shut.

the vessels.[61] Once in the vessels the now “elaborated” fluid was transported upwards. Grew thought of an ingenious mechanism for this transportation, which relied partly on the capillarity of the vessels and partly on the elasticity of the “bladders” (see figure 15). The capillarity of the vessels caused the water to rise a few inches. At the same time the “bladders” surrounding the vessels absorbed fluid from the vessels, which caused them to swell. The swollen “bladders,” pressing on the walls of the vessels, prevented the fluid from flowing downwards. As this mechanism was operative at every successive level of the vessels, the fluid eventually rose to the top of the highest tree, and also from the vessels to the cortex of the stems, because the “bladders” of the insertions passed on the absorbed fluid to their neighboring “bladders.”[62] Grew’s scheme therefore relied on pressure and capillarity as moving agents acting on the fluids in the perpendicular vessels, and suction and pressure as the forces driving the sap from "bladder" to "bladder" in horizontal directions. The appearance of the two principal components of the fabric of plants, the long narrow vessels of the woody parts and the spongelike mass of "bladders" in the parenchymatous parts, conformed perfectly with the action of such mechanical forces.

Within Grew's scheme of nutrition the particles necessary to the plant entered through the root and were conveyed from one "bladder" to another. Although such a scheme implied that the "bladders" have open connections of one kind or another, Grew stated frequently that they are "not visibly pervious" into each other.[63] He therefore assumed a porous structure at a level beyond the resolution of his microscope. The solution of "compounded and heterogenous" principles, once within the cavities of the parenchyma, started to ferment. Fermentation was, in Grew's "doctrine of mixture," one of the manifestations indicating the formation of a new mixture[64] and therefore of new organical substances. The resulting "elaborated" sap was distributed to the several organic parts, a process in which the affinity between particles of the same kind played a major directive role.[65]

On its way through the plant the sap also became "impregnated" or "tinctured" with the main constituents of the tissues it passed through, thus being duly prepared to serve its final purpose. The nourishment and subsequent growth of the parts of the plant resulted "as the *Sap* . . . passeth from *Bladder* to *Bladder,* such *Principles* as are agreeable to those of the *Fibres* of the said *Bladders,* will adhere to, and insinuate themselves into the Body of the *Fibres.*"[66]

When discoursing on the generation of vegetable liquors, Grew pointed out that the nature of the liquid produced depended, as did the qualities of the producing vessels, on both its nature (chemical composition) and the "Structure of the Whole,"[67] by which he meant that the number of adjacent air vessels, for instance, or the position of the said vessel within the total structure to a large extent determined what liquid was produced.

Grew's ideas on the subject of vegetation were therefore composed partly of mechanical and partly of chemical principles. As far as the mechanical operations are concerned, these agreed well enough with the texture of plants as observed through the microscope. Grew's elaborate scheme for the preparation of vegetable substance from the constituents dissolved in the water from the soil also depended on the "Common notions of Sense," that is, on his chemical experiments.[68] Even if this were conceded for the analysis of the "vegetable principles," Grew introduced a conjectural factor into his scheme when he proposed that the position of the various components within the plant body acted as a directive agent in vegetable physiology.

Grew's scientific output was largely confined to botany. Within that subject, however, he tackled a wide range of problems concerning the morphology, anatomy, physiology, and chemical composition of plants. His investigations amply confirmed his original conjecture that plants have a complex structure, which brings about their various physiological processes. In his studies Grew applied two major methods of investigation: observation with the aid of the microscope, and chemical experimentation. Whereas his microscopic research yielded a detailed knowledge of plant structure, Grew derived his notions concerning the physiological processes of the plant mainly from his chemical experiments and subsequent speculations. Past historical studies of Grew's work have been concerned either with his microscopic researches or with his chemical views concerning vegetation. From the present analysis it appears that both were intimately connected.

The anatomical structure of plants, he found, could be reduced to two simple structural elements, the parenchymatous "bladders" and the ligneous vessels. Moreover, these two elements were present in virtually infinite combinations as to bulk, size, and configuration, thus producing the distinctive patterns observed in the various parts of the plants as well as in plants of different species. Grew judged, somewhat too confidently, that his observations of the plant vessels and their structure warranted the conclusion that plant matter, and by analogy animal matter as well, was only constructed from an infinite number of exquisite, extremely fine fibers.

In his physiological conjectures Grew combined chemical principles such as affinity (which in his case was equal to the geometric correspondence between corpuscles) with mechanical concepts such as capillarity, which agreed well with the observed particulars of plant structure. Furthermore, he coupled these concepts with a rather vague notion that the internal arrangement of the plant to some extent determined the qualities of the several component parts. His chemical investigations demonstrated, he judged, that the fibrous nature of organic matter was perfectly compatible with the corpuscular doctrine of matter.

In conclusion, then, Grew arrived at a clear notion concerning the delicate structure of organic matter, which ensued directly from his microscopic studies. However, his ideas concerning the operation of these structures within the vegetable "oeconomy" were partly founded on details, such as minute pores connecting one "bladder" with another, and the extremely fine fabric of the basic vessels, which he was unable to see with his microscope. It is nowhere evident that Grew felt frustrated as a result; this leaves us with the question of why Grew abandoned his microscopic researches. The obvious, and perhaps true, answer is that Grew planned a research program, which he executed to his own satisfaction, and then turned his attention elsewhere. Certainly the five "means" Grew had proposed, viz., an examination of the habitus, anatomy, liquid parts, solid parts, and nutrients of plants, had each been used—some exhaustively, some superficially—to contribute facts and conclusions to his *Anatomy of plants.* Also, after finishing his vegetable investigations, Grew turned to the anatomical study of animals, which only generated a comparative study of guts,[69] and to the description of the Royal Society's collections.

The beginning that Grew made with the examination of plant structure entirely on his own was reinforced by the enthusiastic reaction of the Royal Society. Moreover, the current discussion among the members on the corpuscles from which all matter was considered to be formed encouraged Grew to support his initial findings concerning the fabric of plants with a detailed investigation of their chemical composition, which formed the basis for his views on the construction of vegetable matter from a limited number of different corpuscles. Grew's unique effort to link the fabric of plants with the corpuscular hypothesis was therefore greatly influenced by the ambience of the Royal Society, and therefore associated with both the corpuscular philosophy and the Baconian approach to science.

Vessels Galore

As microscopists studied the animal and vegetable bodies it became apparent that, besides the vascular and fibrous elements known of old, numerous similar but much smaller elements shaped the hitherto amorphous masses of the body. The most conspicuous and universally recurring elements resolved by microscopic analysis of animal and vegetable matter were no doubt the fiber and the capillary blood vessel. As early as 1666 Edmund King published an article in the *Philosophical Transactions* in which he assumed the animal body to be exclusively composed from fibers.[70] Some ten years later Nehemiah Grew followed his suit and explained in detail, as we have discussed in the preceding pages, how the vegetable body was composed from intricately woven fibers. Leeuwenhoek's researches concerning the muscle fibers and the capillary vessels, reinforced by Ruysch's demonstration of the finely wrought capillary vessels throughout the animal body, as well as the microscopic demonstration of delicate fibers in animal tissues by such men as Baglivi, amply substantiated the view that the organic tissues were composed of nothing but vessels, through which circulated the fluids.

The capillary blood vessels were first discovered by Malpighi in 1661, but their distribution through the body and their appearance was much more closely studied by Leeuwenhoek over many years. In 1688 he finally observed how the blood corpuscles are pushed one by one through these tiny vessels with each contraction of the heart. Leeuwenhoek concluded "that the blood vessels which we see in this Animal [the tadpole of a frog], and which we call Arteries and Veins, are just the very same blood vessels; but that they can only be called Arteries so long as they carry the blood into the furthest parts of the small Vessels; and Veins, when they carry the blood back to the heart."[71]

Consequently, Leeuwenhoek's series of observations on the tadpole and his conclusions thereon are heralded by posterity as putting the finishing touch to William Harvey's theory of the circulation of the blood. Leeuwenhoek's contemporaries, however, did not regard his achievement in this light. Even though he frequently asserted that the blood does not flow freely between the fibers of the muscle and other tissues, the idea that the blood from the arteries returned to the veins in precisely this manner was widespread at the time. Indeed, for Baglivi the free flow of blood corpuscles in between the muscle fibers was a necessary prerequisite for his theory of muscle action. Yet the same holds good for those scientists, such as Willis, Croone, and Borelli, adhering to one or other of the "explosion" theories since in such theories the blood and the nervous fluid must both enter the muscle fiber to bring about the required effect. The role of the capillaries within the circulation of the blood was not even the main focus of attention for Leeuwenhoek's colleagues; on the contrary, they discussed the function of the capillaries in relation to other subjects, particularly respiration and secretion.[72] Many different views on their function and how they operated were brought forward and discussed.

Malpighi, in his discussion of the function of the capillaries in the lung, focused on the immense dispersion of the blood as it was spread over the surface of the lung as the crucial factor. At the time of his research it had recently been discovered that digested food is discharged into the bloodstream in the vena cava.[73] From the vena cava the blood is sent to the lungs by way of the heart. In light of this discovery Malpighi suggested that the various components of the blood become thoroughly mixed in the lungs. He thought that the function of the air contained in the lungs was to provide compression of the blood in the vessels during inspiration and exhalation. Thus the lungs act on the various components of the blood in a way reminiscent of the kneading of bread.

In Malpighi's theory the movement of the lungs was indispensable to the proper functioning of the lungs. At this point Malpighi's work was taken up by the English physiologists.[74] Robert Hooke, performing a brilliant but cruel experiment on a live dog, demonstrated that it is not necessary for the life of the animal that the lungs move continually.[75] A continuous supply of fresh air on the other hand is absolutely necessary. This experiment was a sequel to the various air-pump experiments on animals by Boyle and Hooke and their companions in science, demonstrating that the animals died as the air was removed from the cloche of the pump. Another of their group, Richard Lower, demonstrated conclusively that blood changes color in the lungs as a result of its contact with the air.[76] He concluded that blood takes up a certain part of the air during its passage through the lungs, which part he identified with "niter."

A few years later Thomas Willis, Lower's former teacher, studied the anatomy of the lung,[77] which will be discussed in the next chapter. He argued that the network of capillaries in the walls of the vesicles ensured a spreading of the blood over a very large surface—a surface, moreover, which bordered entirely on air. The purpose of this design was, in Willis's view, that the "niter" from the air might all the more easily enter the blood. Willis indicated the "orifices" of the vessels as the means for the niter to enter the blood; that is to say, he imagined that the capillaries of the lung were open-ended. As support for this idea he brought forward the observation that when a fluid is injected into the arteries of the lung, it does not easily pass into the veins as it does in other organs, but a foaming mass appears on the outside of the lung tissue.

Antoni de Heide, who translated Willis's treatise into Dutch, added a note to this passage to the effect that such orifices had not been discovered.[78] On the contrary, all the evidence currently available pointed to the fact that arteries and veins were inseparably connected together by the capillaries, so much so, that it was almost impossible to separate the capillaries into arteries and veins. Indeed, he added, in such capillaries the blood moves now in one direction and now in the other. Here he referred to his own research on the capillaries in the skin of a frog.[79] This research was explicitly allied to Harvey's theory of the circulation of the blood. At the time de Heide wrote it had not yet definitely been established, nor generally accepted, that the blood flowed from the arteries to the veins exclusively through the capillary vessels. De Heide affected not to be interested in the discussions concerning this question, but concentrated on a different problem, that is, on the effects of caustic agents on the blood vessels. He performed his experiments with a live frog on a small strip of its skin flayed from, but still attached to, the body. As an aside to the original effect of caustic fluids (cessation of the flow of the blood) he observed that in neighboring capillaries the blood flowed now in one direction, now in the reverse way. He attributed this "uneven and contradictory flow of the blood" to pathological conditions caused by his experiments,[80] but he also concluded from this experiment that the small capillaries have no valves in them.

The existence of the kind of orifices imagined by Willis was also positively rejected by Leeuwenhoek. Indeed, he had asserted that blood vessels open only into the heart. In the absence of any visible pores, the way in which air entered the capillary vessels remained a problem, and one discussed well into the eighteenth century.[81]

The two mutually opposed explanations for the function of the lungs —that is, the mechanical blending of the constituents of the blood, and the absorption of some portion of the air—both took account of the characteristic properties of the capillary network. There were adherents of both theories in physiological discussion over the next decades. Archibald Pitcairne and Herman Boerhaave, for instance, held to the former position, while Friedrich Hoffman defended the latter. Malpighi also, despite the various criticism and objections brought forward against his conclusions, did not change his views in later years.[82]

The operation of the capillaries was also very much discussed with regard to the secretion of glandular products. Malpighi based his explanation of the operation of the glands on the intimately knotted capillaries and excretory ducts, which he had resolved microscopically. He envisaged a globular body as the basic unit of the glands. Each unit consisted of a follicle with an excretory duct, surrounded by veins and arteries and by branches of the nerves. Malpighi thought that the excretion of the glandular products took place here as a result of the selection of appropriate particles of a definite size and form from the blood. Just how this was effected was beyond his knowledge, he wrote, but he thought that recourse might be had to mechanical similes, such as a sieve. However, the pores of such a sieve were purely hypothetical in Malpighi's view.

Many other writers had recourse to similar simple mechanical operations for the explanation of physiological functions, but were often far less cautious in their theorizing than Malpighi had been. In the Low Countries, for instance, the interaction between pores and particles was a recurrent motif in physiological treatises at the time.[83] Theodoor Craanen, for example, explained in detail how erysipelas and inflammations were caused by particles becoming trapped in the wrong pores.[84] Other writers found in this a reason to challenge this mechanism as a satisfactory explanation. Obviously, if particles of the wrong shape or size might easily obstruct holes in which they became immovably wedged, this constituted a serious danger to the normal operations of the body. Archibald Pitcairne voiced this objection lucidly; he was afraid that "if there is a Necessity for an Agreement of the Pores and the Parts in the Work of Secretion, that no Secretion at all would ever be performed: but since we perceive that frequent and large Secretions are daily and necessarily made in every Animal, we must allow that there is no such thing as that fancied Agreement in the Figures of the Pores, and the Particles secerned, as being what would entirely obstruct the business of Secretion."[85]

Sifting particles through a kind of sieve as a means of selecting the right particles presupposed some kind of selectively permeable barrier between the blood and the final product. The existence of such a barrier was contested by Frederik Ruysch in particular. Ruysch had discovered by means of vascular injections that the blood vessels in all the organs branch out into innumerable small vessels. The path and fragility of these vessels differed from organ to organ. In one organ the blood vessels were arranged in bunches, in others like a brush, and so on. Ruysch judged that the diverse functions of the organs were brought about by the various configurations and contortions of these finest branches. Of secretion he wrote, "I trust that the entire preparation and separation of the humours and fluids, such as saliva, milk, semen, etc., depends only on the quality, variation, and different course of the extremities of the vessels."[86]

Ruysch therefore imagined that the excretory ducts of the glands are continuous with the blood vessels. In this respect Ruysch's views were shared by a number of men, among them Cornelis Bontekoe, who thought that it was not so much the configuration of the ducts as their size that took care of secretion; in his view the fluids became forced into ever-smaller ducts, into which the secretion could pass, but not the larger particles, which were deflected by larger channels branching off the main duct.[87]

The essential difference, whether or not there was a barrier between the bloodstream and the glandular product, could not be resolved at the time. Microscopic research was unable to demonstrate either a direct continuation of a blood vessel into an excretory duct of a gland, or perforations in the coats of the blood vessels. Injection experiments yielded inconclusive results. Some reported the successful passage of the injected fluid from the arteries into the veins or even excretory ducts,[88] whereas others failed.[89] Therefore, neither theory could be demonstrated conclusively, so that preference for a particular theory could only be the result of philosophical commitment or personal sentiments. Malpighi's fundamental bias was toward purely mechanical actions, so that the various sizes and shapes of the pores and particles appeared to him to constitute the most likely working principle in the case of secretion. To Ruysch, on the other hand, secretion seemed to be the result of the shape and course of the duct conveying the blood. This view was encouraged by the numerous exquisite demonstrations of so many different variations of this by means of the injection technique in which he excelled.

The rivalry between Ruysch and Malpighi was to some extent the result of the technique chosen for the investigation of animal organs. Through his microscope Malpighi saw globular structures, for instance, the glomeruli of the kidney, and—starting from there—arrived at a globular body. Ruysch on the other hand saw that the ends of the vessels into which he had injected some sort of material fade away into ever greater thinness. However, he added, one did not need a microscope to see what was demonstrated so plainly for all to see in his anatomical preparations,[90] ignoring the fact that he and Malpighi therefore observed vessels of a completely different size.

Ruysch's conception of the operations of the body, the *oeconomia animalis,* depended entirely on the course of the capillary vessels, as opposed to Malpighi's notion of mechanical actions. Contemporary physiological theorists, aware of the fact that the microscope, as well as injection techniques, revealed that the solid parts of the body were invariably composed of lesser parts similar to the larger—that is, vessels were composed of smaller vessels—muscles of smaller muscle fibers, and nerves of even smaller nerve fibers, gradually favored the capillaries as the seat of all physiological processes. By then the relevant anatomical facts concerning the main transportation systems for fluids throughout the human and animal bodies, that is, the blood vessels and lymphaducts, had been adequately documented. Consequently, the physiologists explored the conduct of the fluids within the vessels with mathematical tools, rather than with instruments, experiments, and observation.

Influenced by the recent developments in physics a "Mathematical Physick" was introduced and physiological theory shifted from "machinulae," or mechanical actions, to the dynamics of fluids in tubes of various sizes. Archibald Pitcairne,[91] who was deeply influenced by Newton's philosophy and analytical methods, developed a physiological system of this kind, in which all the operations of the body depended on the hydraulic effects within the vessels of various sizes. Hydraulic physiology, in several guises, dominated eighteenth-century physiology.

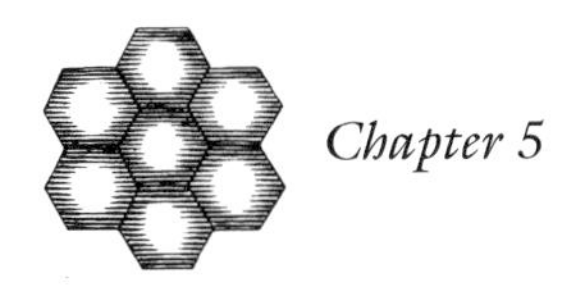

Chapter 5

The Fabric of Living Beings

IN THE PRECEDING chapters we have examined microscopists who used their microscopic findings to substantiate the mechanical philosophy and microscopists who unreservedly interpreted their microscopic findings within the context of a mechanical explanation of bodily functions. However, the microscope was also employed by investigators who were simply concerned with elaborating traditional anatomy and natural history, that is, the description of the habitus and habitat of animals and plants. The Baconian appreciation of amassing data certainly gave encouragement to the efforts of the naturalists engaged in cataloging the various kinds of organisms, and it encouraged anatomists to expand their explorations of animal form. The activities of the seventeenth-century naturalists and anatomists were generally speaking not very revolutionary insofar as method and subject matter were concerned. However, an optical device that greatly enhanced the powers of vision could hardly fail to meet with an enthusiastic reception among those interested in the accumulation of data concerning the living world. Accordingly numerous anatomists and naturalists of the late seventeenth and eighteenth centuries used a microscope or a magnifying glass to study the details of their specimens.

Animal and Human Anatomy

The second half of the seventeenth century was a flourishing period in anatomy. It has been described as "prolific in detailed anatomical observation."[1] The collective efforts of anatomists throughout Europe yielded numerous studies in which particular organs or physiological systems were carefully described, with a wealth of newly discovered details such as that of the lymphatic system (Thomas Bartholin, 1653), the brain (Willis, 1664), the muscles (Stensen, 1666–67), and the male and female genital organs (De Graaf, 1668 and 1672 respectively). The contributions made by microscopic research to the expanding corpus of identified and characterized particulars of the human body included the capillary blood vessels, the corpuscles of the blood, and the spermatozoa. Within a short span of time it was established that the various organs, and other structural parts of the body such as bones, consisted of an intricate arrangement of very fine elements. Certain elements, such as the capillary blood vessels, were found to occur in all organs, but others appeared to be characteristic of certain organs such as the reticular tunic covering the sensory papillae of the tongue.

In this period much attention was also paid to the animal body, usually with the express desire of understanding the human body better. These studies were consequently of a comparative nature. For this purpose a varied array of hitherto little-studied animals were studied, including many species toward the lower end of the scale of the Great Chain of Being. At the same time the anatomy of plants began to be seriously studied, as we have seen in the previous chapter, by Malpighi, Grew, and Leeuwenhoek. Although lengthy discussions were held at both the Royal Society and the Académie des Sciences, and various experiments were performed concerning the question of how and from whence liquids were transported through the plant, no further investigations into the microscopic anatomy of the plant were undertaken.[2]

Anatomical research in the seventeenth century and after was to a large extent subservient to the investigation of the functions of the human body. Precise knowledge of structure formed the basis on which physiological theorizing was founded. A large number of such investigations were undertaken with animal specimens, and relevant anatomical data were often, albeit somewhat incautiously, transposed to the texture of the human organs. A case in point is the structure of the lung. Malpighi's knowledge of the structure of the lung was founded on the dissection of a dog, a frog, and a tortoise, but he applied the results specifically to the human frame. Similarly, Thomas Willis based his conclusions on the structure of the human lungs on his research with dogs, sheep, and oxen. This research, which was executed with the help of Edmund King,[3] was published in 1674 and 1675 in the two volumes of his *Pharmaceutice rationalis.* In this book Willis attempted to explain the actions of various drugs on the human body, and his first concern was to elucidate how and in what ways drugs reached the "spirits" residing within the human body. He therefore began his treatise with an enquiry into the finer details of the anatomy of the alimentary duct, since this was the principal route through which drugs entered the body, and complemented this with an anatomical study of the respiratory organs.

It is not absolutely clear whether his examination of the stomach and gut was indeed executed with the help of the microscope, but if so, it was certainly carried out with small magnification. Nevertheless, Willis's practice was to use the microscope when this was relevant. The investigations concerning the lung, reported in the second part of *Pharmaceutice rationalis,* certainly did benefit from microscopic examination. Willis was conversant with Malpighi's writings on the subject, published some twelve years earlier, and elaborated on them; indeed, he wrote, he had "searched out and depicted everything about the chest and lung, or searching through the most secret parts, everything that ancients and moderns have written, and everything that may be discovered with the dissection knife and through magnifying glasses."[4]

Willis, in complete agreement with Malpighi's findings, found that each lobe of the lung is subdivided into smaller lobes and that the latter are attached like grapes along the length of a bronchus. The outer wall of each of the smaller lobes appeared to be subdivided into the vesicles, first discovered and described by Malpighi. Willis also investigated the membrane covering the lungs, and noted the imprint of the lung vesicles on this tunic. He found that the nerves as well as the veins and arteries branch out in the same pattern as that of the trachea. One of his original contributions to the anatomy of the lung was his discovery of the lymphatic vessels on the outside of, and running through, the lungs. He thought that these vessels originated in the spaces between the smaller lobes of the lung. He found the trachea to be composed of four different coats, an inner muscular coat and an outer cartilagineous one, separated by a glandular and a vascular tunic. The walls of the bronchi appeared to be fitted with circular and longitudinal muscular fibers as well as cartilagineous rings, and the larger blood vessels were found to be composed of four separate coats. In a series of clear illustrations Willis depicted the various details in the anatomy of the lobes of the lung, trachea, and large blood vessels in the lung.

Where the alimentary duct was concerned, Willis and King concentrated their research on the structure of the walls of the esophagus, stomach, and different parts of the intestines. In the preface to his book Willis wrote that they had "clearly discovered in the stomach and gut, without the direction or help of other writers, the nervous, fleshy and glandular coats, also some sensitive and other movement fibers, as well as very dense networks of blood vessels and numerous glands."[5] They found that the different walls were each fabricated from several separate tunicae. In the esophagus and stomach they distinguished three layers, and in the intestines four. Willis described these various layers and the differences between them in minute detail, and associated the varieties in the structure to the diverse actions of specific drugs.

Apart from Willis few anatomists contributed original microscopic research on human materials. The investigations of the Spanish anatomist Crisostomo Martinez are an exception rather than the rule. In connection with his research concerning ossification, he prepared several drawings of the structure of human bones based on original microscopic research executed ca. 1685 (at which time Martinez often resided in Paris), the details of which were, however, not published until the present time.[6]

Even if many of the anatomical details mentioned by such men as Willis were only discerned in animals, they still found their way into such unadulterated anatomical publications as the anatomical atlas. One example of a late-seventeenth-century atlas is of considerable interest to the present enquiry, as it was the first atlas to incorporate most of the recently discovered microscopic detail. This was Goverd Bidloo's *Anatomia humani corporis,* published in 1685.[7]

Bidloo attempted to represent, besides the traditional subject matter of an anatomical atlas, much of the knowledge then available on the microstructure of the human body. For that part of his atlas his obviously very sensible plan was to make exhaustive use of a number of recently published treatises, mainly those by Malpighi and Willis, as a guide. He incorporated illustrations of the microscopic structure of the skin, tongue, liver, kidney, spleen, brain, lung, and trachea, the wall of the stomach and blood vessels, and several other structures. As Malpighi's treatises on the various organs were not illustrated except for the second and subsequent editions of his *De pulmonibus,* Bidloo was obliged to repeat most of Malpighi's observations in order to plan and execute his illustrations. In fact he claimed that he had dissected and studied all the material depicted, writing, "I show nothing, I repeat, nothing, after the illustration of others, I hate the slavish labor of extraction."[8] Insofar as the macroscopic part of his anatomies is concerned, it is evident that Bidloo personally prepared and dissected bodies to be drawn by his fellow worker in this particular enterprise, the artist Gerard de Lairesse. A critical analysis of Bidloo's microscopic illustrations demonstrates that, although the author had actual specimens portrayed in most cases, some of his illustrations either do not show the details described or display these in a very artificial, contrived manner.[9] On one side of the dividing line stand the rather large structures, such as the papillae of the skin and tongue, and those that had already been illustrated by others, such as the vesicles of the lung. On the other side stand the much smaller structures, such as the follicles of the spleen. This suggests that Bidloo encountered serious problems when trying to verify the more minute microscopic details, particularly in those cases where there was no illustration available in print to guide him.

Bidloo has been accused of concocting his illustrations from Malpighi's published descriptions (his representations of the microscopic structure of the kidney and the brain are cases in point) rather than drawing from actual specimens.[10] This allegation suggests that Bidloo in fact failed to distinguish such minute or illusive structures as the follicles in the spleen and the "glands" in the brain. However, on Malpighi's authority, Bidloo in these cases simply constructed an illustration containing all the details to be gleaned from Malpighi's description. The fact that Bidloo felt no compunction about resorting to such a procedure is substantiated by his letter on the liver fluke, which he addressed to Leeuwenhoek and published in 1698, entitled *Brief aan Antony van Leeuwenhoek wegens de dieren, welke men somtyds in de lever der Schaapen en andere beesten vind* (Letter to Antoni van Leeuwenhoek concerning the animals that are sometimes found in the liver of sheep and other animals). The portrayal of the liver fluke is very acceptable except for the two eyes with which Bidloo equipped them and that these organisms in fact lack. To make matters worse, he included an enlargement of these eyes in which they are reminiscent of fish eyes. Bidloo was convinced that all animals must possess eyes, even though they are not readily detected, a point he substantiated many years later with an account of his dissections of a mole and a kind of snake, which were commonly regarded as eyeless animals.[11] As he felt sure that a liver fluke should have eyes like any ordinary animal, he did not hesitate to include these.

It seems likely that in the case of his anatomical atlas an equally strong belief in the verisimilitude of Malpighi's findings prompted the abovementioned clashes between text and illustration. The microscopic structure of the spleen, for instance, is described as being conformable to Malpighi's report in *De liene,* but the accompanying illustration shows none of the details mentioned.[12] Some patches of shading in the illustration served to disguise the fact that Bidloo had failed to observe these. In the case of the

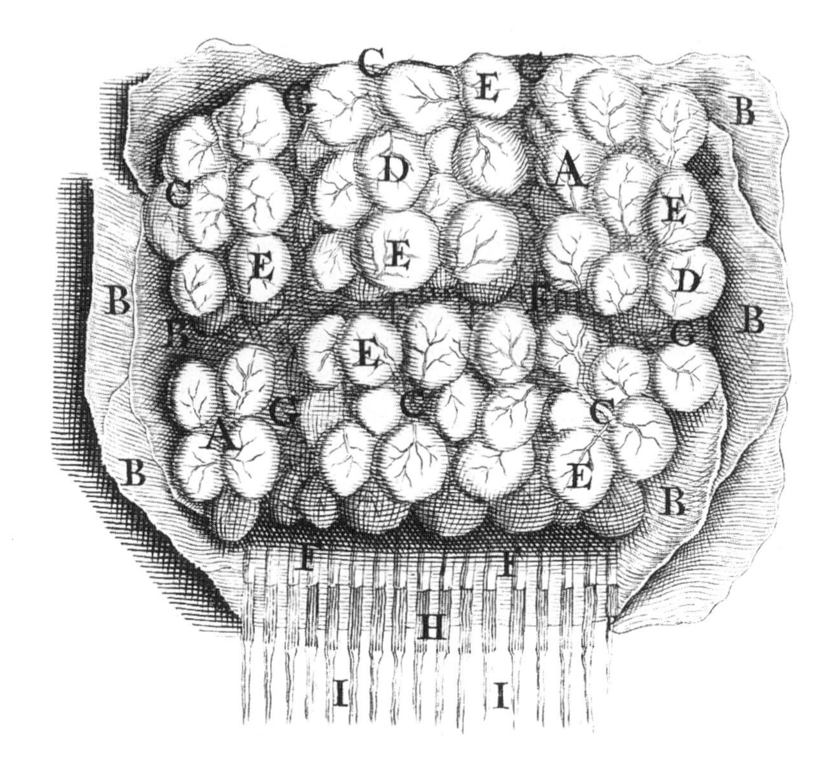

Fig. 16. The fabric of the brain according to Malpighi, as represented by Bidloo in table 10 of his *Anatomia humani corporis* (1685).

brain (see figure 16) Bidloo presented a complete fabrication comprising clusters of spheres, which stood for the "brain glands" and which were covered with capillaries. The arrangement of the glands into clusters was presumably inspired by the spherical surface of the gyri, but Malpighi certainly had not envisaged the glands as being of the size depicted by Bidloo.

Despite its shortcomings, Bidloo's atlas constitutes a landmark in human microscopic anatomy. The majority of his illustrations of the microscopic details of human organs were based on his own observations, and about half of these structures had not been depicted in print before. Moreover Bidloo attempted, rather successfully, to capture the three-dimensional structure of most of his objects. Certainly Bidloo's endeavor was not matched by any of his contemporaries. On the contrary, some of his illustrations were copied by contemporary writers of anatomical textbooks, such as Steven Blankaart in the second and subsequent editions of his *De Nieuw hervormde anatomie* (1686ff.) and Philip Verheyen in his *Anatomiae corporis humani* (1693).

In 1681 Gerard Blaes published a survey of contemporary studies of animal anatomy, entitled *Anatome animalium*. This work was in fact a compilation of recently published observations by a range of authors, drawing heavily on the works of Niels Stensen, Thomas Willis, Thomas Bartholin, and Regnier de Graaf. Blaes also included a great deal of the data collected by the Amsterdam anatomical society, the Collegium Privatum, of which he and Swammerdam were the principal members. In his book Blaes dealt with terrestrial animals ranging from the chameleon to the camel, flying and aquatic animals, and several insects. The majority of the animals were also described and portrayed, for which purpose the illustrations from the original books had been copied. Comparative anatomy, as emerges from Blaes's book, had become a richly documented subject by 1680. Many animal species had been thoroughly studied, although most of the zootomies do not appear to have benefited greatly from microscopic examination.

The same may be said of the dissections reported in two small booklets entitled *Observationes anatomicae selectiores* (1667) and *Observationum anatomicorum collegii privati Amstelodamensis pars altera* (1673), published by the Collegium Privatum of Amsterdam. This body of men studied a number of animals, particularly fishes, and only occasionally resolved structures with the microscope, even though Swammerdam took a major part in the execution of the dissections. An exception to the rule was the microscopic study of the internal appearance of the halibut pancreas.[13]

It seems, then, that as long as the animals examined were of noticeable size, traditional means sufficed to study their anatomy. Insofar as the microscope was employed it served to resolve poorly visible details more clearly. The same approach is apparent in the exploits of a group of French anatomists led by Claude Perrault, who all performed numerous dissections of a series of animals between 1667 and 1676. The results of these studies were published as *Mémoires pour servir à l'histoire naturelle des animaux*. In the preface to this work it is clearly stated that one or other of the coworkers often inspected the parts they had just dissected through the microscope. Despite this assertion the impressive illustrations and the descriptions include hardly any microscopic detail; the portrayal of the microscopic appearance of the various tunicae in the gazelle stomach is one instance of the few.[14] Perrault's objective was to examine the anatomy of obscure animals thoroughly, and he conceived a particular interest in the explanation for the mechanism of the unusual anatomical peculiarities he found in the organs of some of these animals. Cases in point are the tongue of the chameleon and the spiral gut of the shark.

Perrault's English counterpart, Edward Tyson, on the other hand found much more use for the microscope, mainly because he occasionally directed his attention toward such inconspicuous animals as parasitic worms.[15] Tyson advocated the comparative method in anatomy in the prologue to his famous *Phocaena, or the anatomy of a porpess* of 1680, as the royal road to a complete natural history of animals. As he was a firm believer in the Great Chain of Being he expected that, in the course of preparing such a natural history, one species often offered a key to the understanding of another. Moreover, anatomical research would also lead to the intimate knowledge of the operation of nature within living beings, the animal "oeconomy." Indeed, he wrote, "Nature's *Synthetic* Method in the composure and structure of Animal Bodies, is best learnt by this *Analytic,* by taking to pieces this *Automaton,* and viewing as under the several Parts, Wheels and Springs that give it life and motion."[16] From these sentences it appears that Tyson shared the hopes of a number of his fellow members of the Royal Society, who presumed that the microscope would provide a material basis for the mechanical explanation of natural phenomena.

Tyson's scheme for the execution of the natural history of animals included four lines of approach: description of individual specimens; analysis of the fluids of the body; comparative description of the parts of animals; and *embryotomia,* or a study of the development of the animal body. A regular role was reserved for the microscope in providing a complete and accurate description of the solid parts of the body. Tyson made regular use of the microscope throughout his work and used the instrument in general to elucidate the smaller structures of his specimens, like the Parisians. Yet as Tyson examined a number of imperfect animals, unlike the Parisians, he was able with the help of the microscope to resolve the mysteries concerning some problematical species, for instance, the true nature of the cochineal,[17] the mode of reproduction in a species of parasitic worms,[18] and the establishment of the animal nature of hydatids.[19] Tyson's fame in anatomy rests securely on his examinations of the porpoise, rattlesnake, opossum, and chimpanzee, but he also studied some invertebrates, a roundworm and a tapeworm, in connection with the question of spontaneous generation. By that time, 1683, the studies by Redi and Swammerdam had definitely settled this question as far as insects were concerned, but the parasitic worms remained problematical and continued to be so well into the nineteenth century.

Dissecting the roundworm (see figure 17) Tyson found that the male and female reproductive organs are easily laid bare to the microscope-aided eye, and he was also able to identify the eggs. Therefore, the question of whether roundworms reproduced sexually could be answered in the affirmative, but in the case of the tapeworm Tyson was less successful. He observed a spiked "head" (in fact the rostellum), to which were attached a large string of nodes, becoming successively larger toward the tail of the organism. He also noted protuberances (the genital papillae) at the side of each of these nodes. He carefully examined these, but the only use he could imagine for them was that of a mouth. He did not dissect this organism and therefore found no sign of reproductive organs.[20]

Whereas Perrault and Tyson were engaged in successive, wide-ranging anatomical investigations spread over many years, a fairly large number of contributions to comparative anatomy were made by men who published only a few anatomical studies. One of these contributors was Antoni de Heide,[21] a doctor from Middelburg, who studied among other things the mussel, a common foodstuff from the seas surrounding the area where he lived. In 1683 de Heide published a short treatise on this animal, entitled *Anatome mytuli*, in which he described its anatomical details. From these details it is clear that de Heide used a microscope to good effect. Especially noteworthy is his description of the ciliated epithelium (see figure 18), which he recognized particularly on the surface of the gills by its conspicuous "motus radiosus seu tremulus" (a radiant or tremulous motion). He wondered how this motion was brought about, and at first assumed that it resulted from a continuous stream of some kind of fluid circulating over the surface of the gills and other parts of the body. However, he rejected any hypothesis of this kind as he observed that the filaments, comprising the labial palps, moved; moreover, they moved in all directions and were not bent, like a tree in the wind, all to one side. Finally, as he severed a part of a labial palp from the body, he observed that the motion did not cease for some two hours, a fact that could not be reconciled with a circulating fluid. De Heide therefore decided that the radiant or tremulous motion on the surfaces of the parts of the mussel was brought about by the movement of the minuscule filaments, the cilia, on these surfaces.[22]

Although anatomists declared the microscope to be a valuable contribution to their arsenal of techniques, and valued the microscopic examination of specimens, in fact few anatomists attempted to investigate the

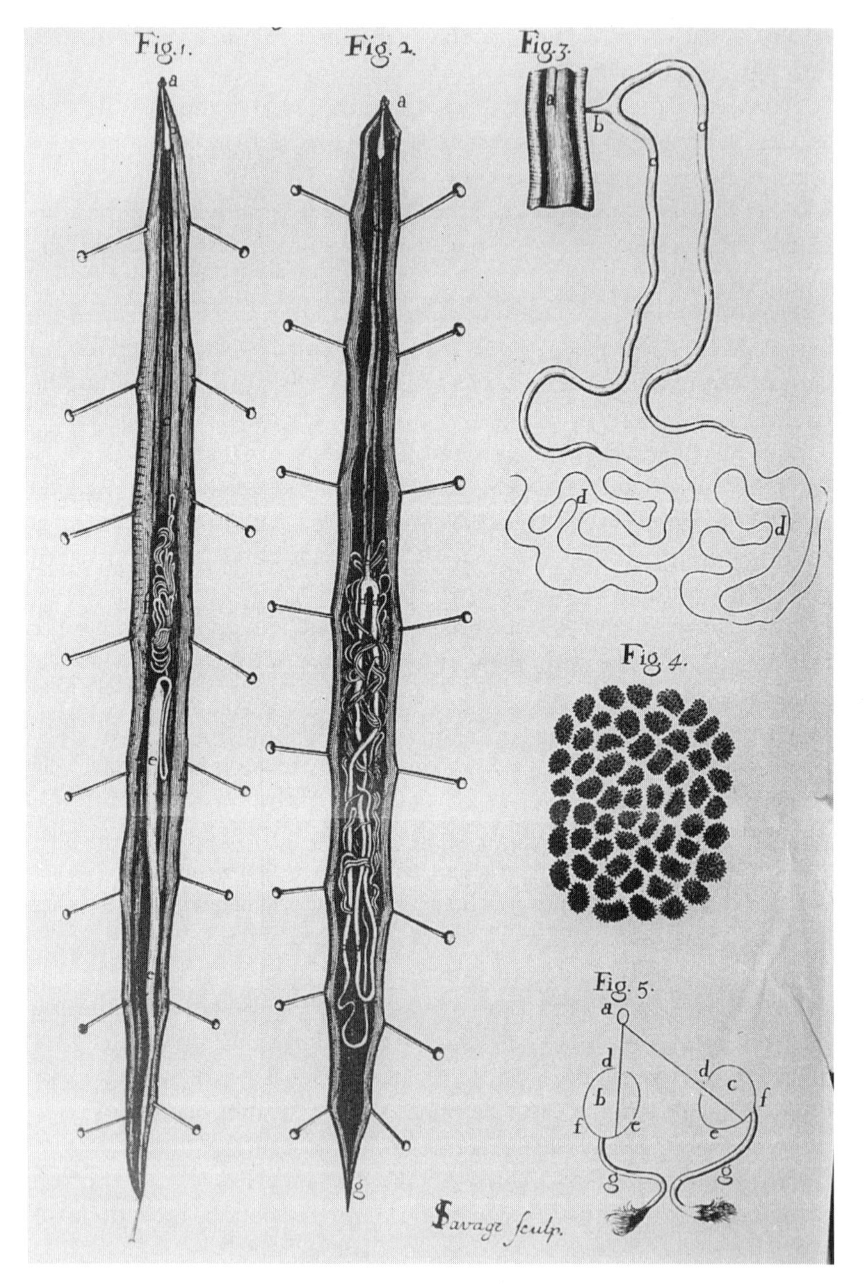

Fig. 17. The anatomy of the roundworm as delineated by Tyson in an illustration to his article in the *Philosophical Transactions* (1683).

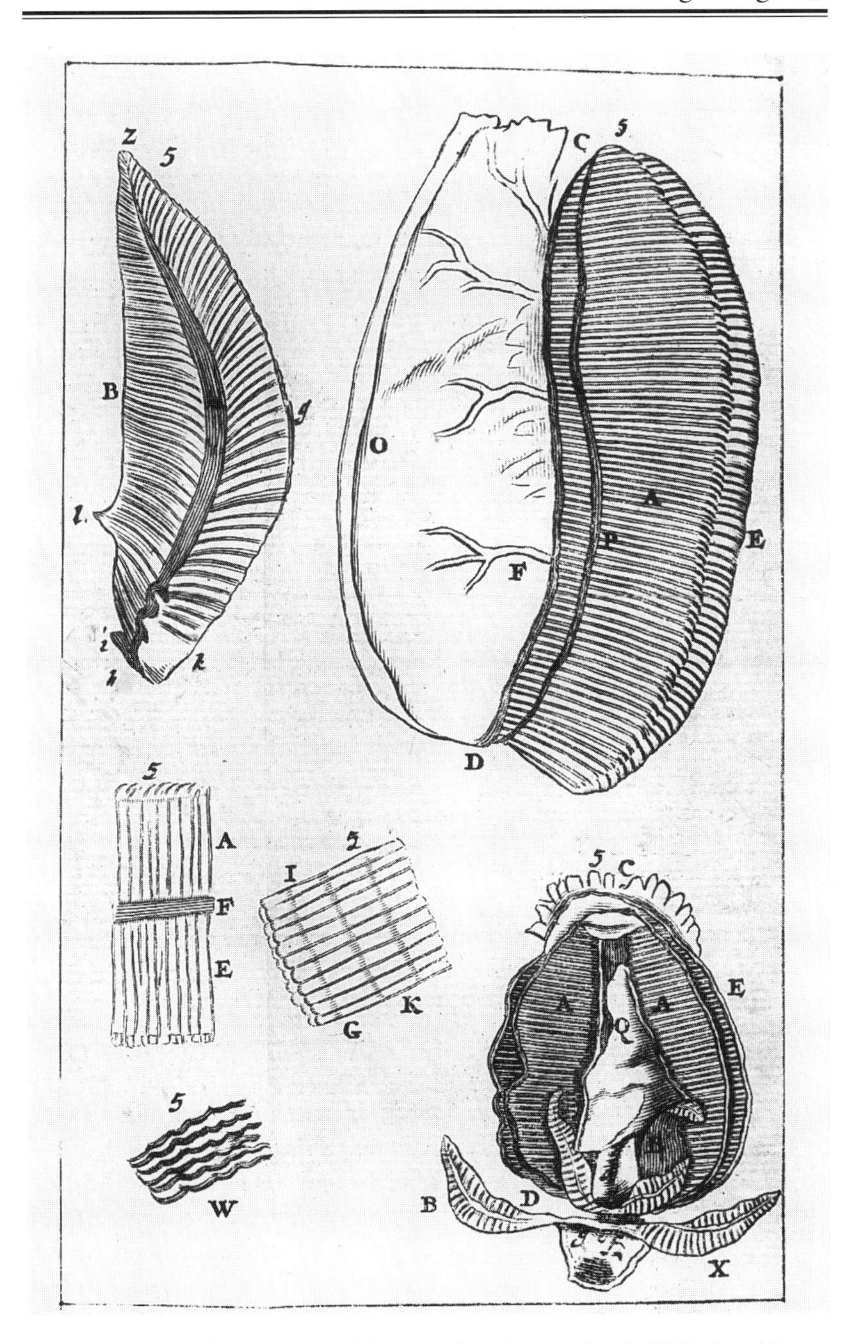

Fig. 18. Details of the anatomy of the mussel as depicted by de Heide in table 5 in his *Anatome mytuli* (1683).

texture of human organs with the microscope thoroughly, neither directly nor via the roundabout route of comparative anatomy. The investigations performed by Malpighi and Leeuwenhoek were, as far as human anatomy is concerned, only expanded through the explorations of Willis, Martinez, and Bidloo. However, Bidloo's endeavor, impressive as it may be, can hardly be rated as an original contribution to human anatomy as far as the choice and elaboration of his subjects is concerned. The evident flaws in Bidloo's atlas only serve to underline the fact that some of the superior findings of the leading microscopists, particularly in regard to the texture of the organs, could not easily be verified by their contemporaries.

The study of the animal body, on the other hand, was certainly carried some way beyond the achievements of the "fathers of microscopy," but this achievement was one of quantity rather than of quality. Investigators like Perrault *cum suis*, Tyson, and de Heide anatomized many unknown species and consequently contributed substantially to contemporary knowledge of animal anatomy. Their efforts revealed that, even though the familiar plan of construction of the animal and human body was not common to all animal species, nevertheless comparable structures were present in all, or almost all, species so far studied.

Jan Swammerdam

Swammerdam did not often apply the microscope to the anatomy of the human and large vertebrate body. In his *Miraculum naturae,* for instance, which deals with a subject that could benefit from the application of the microscope, that is, the human ovary, he does not mention its use. Nor were other suitable subjects, such as the medulla spinalis, microscopically investigated. The research published by the Collegium Privatum of Amsterdam,[23] which primarily concerns larger vertebrates, especially fishes, is only infrequently augmented with microscopic details. There was one topic that Swammerdam proposed to study experimentally with the microscope; this was the question of whether blood also contained the globules that had been observed in samples outside the body, as it coursed through the body. He proposed to insert a tube into a dog's vein, guide it past a microscopic apparatus, and lead the blood back into the body without ever exposing it to the air.[24] This experiment, which Swammerdam devised in 1678, was never actually performed.

In his *Bloedeloose dierkens* Swammerdam mentioned the advantages offered by the microscope, but from the content of this book it appears that his close scrutiny of the insects did not involve the application of any optical instruments other than rather weak magnifying glasses. His *De ephemeri vita* of 1675, however, and even more so the *Biblia naturae,* abound with magnificent microscopic studies. It is therefore clear that Swammerdam only really concentrated on microscopic research from approximately 1670. However, he had acquired sufficient microscopic technique some years before to state, in his doctoral thesis, his views on the construction and operation of the organ of sight in insects.[25]

Swammerdam learned from Johannes Hudde how to produce lenses for single microscopes some time in the 1660s.[26] These were the small globular blown lenses made from melted glass, a technique that was very simple, enabling Swammerdam, as already mentioned in chapter 1, to make over forty of these lenses—some bad, some good—in an hour.

Swammerdam's greatest asset was his mastery of microscopic technique. With great perseverance and ingenuity he endeavored to dissect the minute bodies of some twenty different kinds of insects, among them a louse, bee, and flea, in various stages of their development. To do so he used the conventional tools of the anatomist: knife, pincers, scissors (to which he was particularly partial), and needles, all of which were of course of delicate make. Some of the tools he acquired with considerable difficulty in France through Thévenot.[27] Over the years Swammerdam developed various techniques for improving contrast in his preparations, the lack of which he found to be one of the main obstacles in establishing the details of the insect anatomy.[28] He injected wax, tin, or colored liquids into the vessels, or simply used colored glass for a background, but he also applied coloring agents and dried his preparations, to mention but a few of the techniques to which he referred in his writings. Swammerdam had probably mastered quite a few more besides. He once wrote proudly to Thévenot that he had developed numerous new techniques, which stood at the basis of his various discoveries.[29]

Moreover, he was well aware of the limitations and deceptions of lenses, however much he valued their powers. Swammerdam warned repeatedly that interpretation of the image should be carried out with care because it is not always easy to discern, for instance, between a hollow or spherical surface.[30] In his drawings, and perforce also in the published illustrations, Swammerdam chose not to depict all the details according to life, "because I think that too sorry an exertion and of little value,"[31] but preferred to depict the more important things as being slightly bigger than others.

The main emphasis of Swammerdam's microscopic research is on the anatomical structure of the insect body in its various developmental stages. Thus he noted the main features of insect anatomy, the general plan of insects' internal arrangement, experimenting at times with these parts so as to elucidate their operation. One instance of his thoroughness is provided by his experiments with the venom bladder in the bee and the adjoining stinging apparatus. He made the bees sting a wash-leather glove, and collected some of the venom to taste it in order to determine its nature.[32] Exploring the structure of the insect body Swammerdam could not but conclude that the structures he observed were essentially the same as he had seen before in larger animals.[33]

As an example of the care and fineness of detail of Swammerdam's microscopic research, his dissection of the compound eye of the bee can be cited (see figure 19). The eye of an insect, as he had already remarked in his doctoral thesis, has a reticulated outer surface. This is most easily observed when this surface, or horny layer, is separated from the rest of the eye and seen against the light. Subsequent research revealed the complex structure of the compound eye; beneath the outer layer, "There are so many fibers as the horny layer of the eye on top has divisions: these fibers enclose quite nicely the bulging of the spherical divisions of the horny layer. Their form on top is six-cornered and wide, thinner in the middle, and pointed at the end, moreover they are all of the same length, thickness, width, and magnitude."[34]

Between these pyramidical fibers he noted numerous air vessels and found that the points of the pyramids rested on a double-layered membrane, covering a brainlike substance. The outer surface of the membrane had a dented appearance caused by the imprint of the tops of the pyramids. The brainlike substance was described by Swammerdam as "another or second kind of fibers, which are laid transversely against the undersurface of the described membranes, and appear as the foundation beams of the surmounted pyramidical fibers. These fibers differ from the upper pyramidical [ones] in that they are not so numerous as those and also by far not as delicate."[35] Judging by their appearance these fibers constituted the cortex of the brain; this opinion was supported by the fact that Swammerdam, in contrast with Odierna, observed a link between this substance and the brain proper.

Swammerdam saw that the bases of the pyramids contained a colored substance. This substance usually came away together with the horny layer when this was separated from the rest of the eyes. Swammerdam identified this substance with the retina. Therefore, according to Swammerdam, the

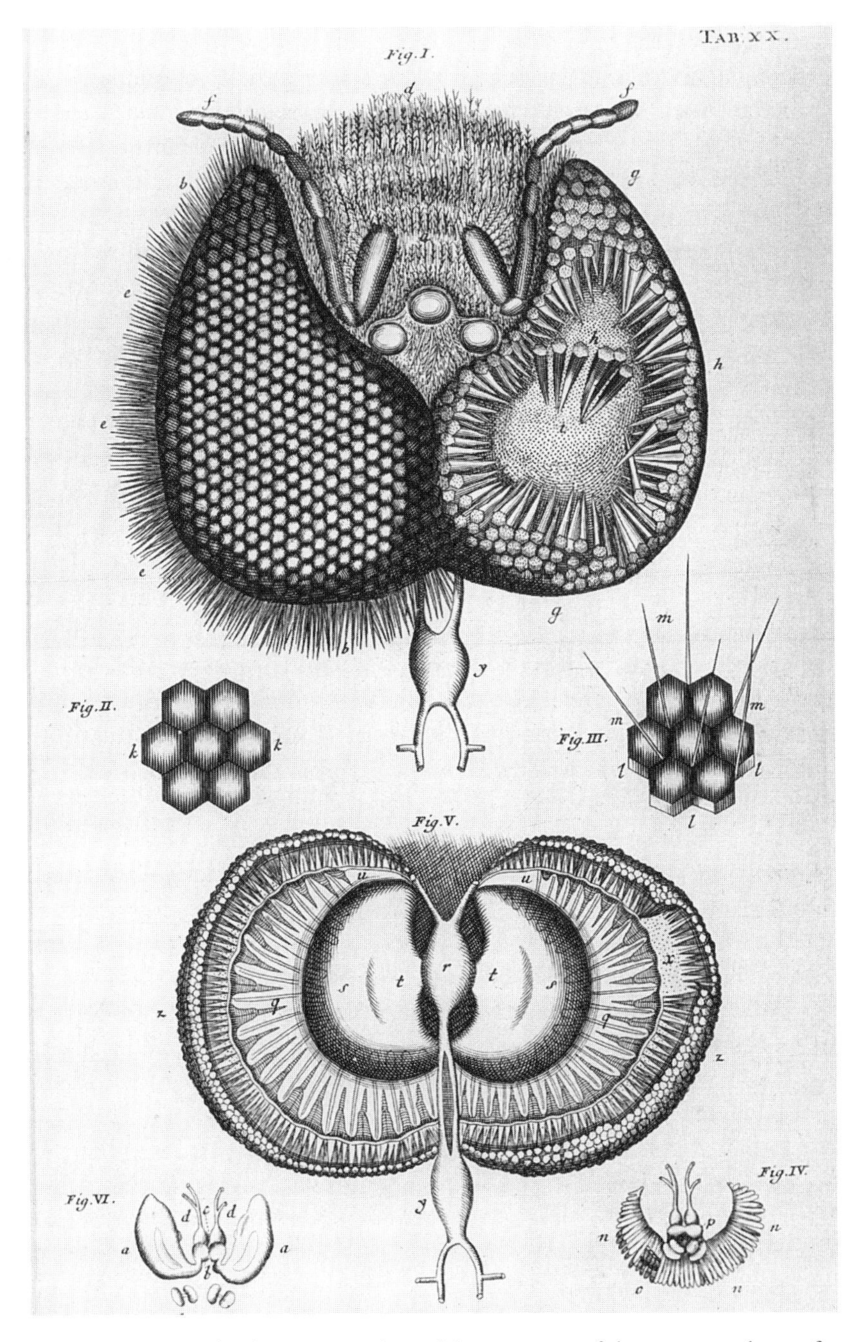

Fig. 19. Swammerdam's representation of the structure of the compound eye of the bee, table 20 in his *Biblia naturae* (1737–38).

retina of the insect eye was located directly beneath the horny layer. Swammerdam's ideas concerning the operation of the compound eye are shaped by this proposition. He wrote, "These eyes are thus constructed in such a way that they receive the images of things by a single propulsion of the reflected light."[36] In *De ephemeri vita,* Swammerdam had elaborated this idea somewhat, writing, "The vision of these animals [insects] operates in a quite different manner to that in us, where it comes about through a gathering of rays within the eye. In these it comes about by a gathering of nervous fibers, which, at the moment of seeing, are but shortly touched and moved on their bulging by the visible qualities and the rays of light and color."[37]

Swammerdam therefore thought that images were formed directly beneath the outer surface of the compound eye, and that a multitude of images were separately relayed through the fibers to the brain. Although Swammerdam's description of the construction of the compound eye was by far the most detailed of any of his contemporaries, his explanation of the operation of this eye is rather unsatisfactory, particularly so in light of the very lucid account of the optical properties of the compound eye presented by Hooke in his *Micrographia,* which Swammerdam knew well. However, he criticized Hooke's reference to a fluid within the eye which, as Swammerdam wrote somewhat accusingly, Hooke needed to explain the operation of sight, but which in Swammerdam's experience did not exist.

Swammerdam's research on insects was concerned with their development, appearance, anatomy, and way of life. Swammerdam described minutely the details of the alimentary duct, the respiratory, circulatory, and nervous systems, as well as the generative organs. In so doing he was guided by the general plan of the vertebrate body. For instance, having localized the stomach he could work out the various other parts of the alimentary duct, keeping in mind the succession of parts in the vertebrate body. Moreover, he was conversant with the data of contemporary investigators, particularly those of Hooke and Malpighi, both of whom he praised highly.[38]

In his *Biblia naturae* Swammerdam described the details of the appearance, development, and anatomy of insects in a breathtaking way. From an examination of its contents it appears that throughout the preparation of this impressive work he kept strictly to the rule that observation and experiment must form the basis for subsequent theorizing. In Swammerdam's opinion, the object of scientific enquiry was to explore the details of God's creation and to study and describe the phenomena of nature, whether they be the movements of the stars or the anatomy of a louse. With the latter object in view the microscope served as an indispensable tool with which to resolve the finest details of the anatomy of insects and other minute creatures.

The exacting attitude to scientific investigation that induced Swammerdam to concentrate on meticulous description was strengthened during the mid-1660s, when he worked more or less simultaneously on the problems of respiration in vertebrates and the metamorphosis of insects. In keeping with the tenets of empirical science he tackled the problem of respiration by means of a series of experiments. The explanations he subsequently advanced were aimed to fit in with the prevailing mechanistic explanatory framework. However, Swammerdam realized that mechanistic explanations for physiological processes, such as his own explanation for respiration, were inadequate and artificial, particularly so since at the same time he had come to see, in the course of his entomological investigations, that careful observation led to new and original results, results that he deemed important and of which he was proud.

Unlike Hooke and Malpighi, Swammerdam did not harbor any preconceived notions on the structure and functioning of the living body that might be clarified by a closer look at the details of the anatomy of organisms. He was content, certainly as far as his microscopic investigations were concerned, with recording nature, and the only kind of explanation he attempted was teleological. He presumed a close analogy between the function of parts in insects and the function of parts similarly situated in the bodies of higher animals. Only in a very few instances, his views on the operation of the compound eye being a case in point, did Swammerdam propose a manner of operation.

From numerous statements in Swammerdam's *Biblia naturae* it is apparent that the main motivation for his microscopic study of insect anatomy stemmed from his deeply felt admiration for God's magnificence. In his view God's guiding hand was nowhere more manifest than in the orderly and intricate arrangement of processes and structures in organic nature. The observation of the delicate and exquisite fabrics of living beings, revealed when he began to use a microscope to explore the anatomy of insects, could only support his original point of view. Moreover, by the time that Swammerdam was able to complete his studies on insects (the second half of the 1670s) he had resolved the earlier conflict between traditional forms of worship and his passionate commitment to science. By that time he regarded the description of nature as a form of worship.

The Anatomy of Insects

Insects proved especially rewarding objects for microscopic inspection. The investigations of insects may be divided into two categories: the description of habitus and anatomy. The habitus of small insects was studied microscopically by many men ranging from Borel to Hooke and Giacinto Cestoni to Leeuwenhoek. Apart from detailed descriptions of the exterior parts of a wide range of unexplored species, a host of interesting facts concerning insects became known through their research. Examples are the parthenogenetic reproduction of aphids,[39] the nature of the cochineal insect,[40] and the pathogenic effects of some minute parasitic insects.[41]

The examination of the anatomy of insects obviously required much more skill in the microscopist than the description of its appearance. In fact, the three pioneers in this field—Malpighi, Swammerdam, and Leeuwenhoek—were unparalleled in their lifetimes. Through their investigations it was established that the familiar features in the plan of construction for mammals, fishes, and reptiles could also be found in the smallest of insects: the digestive tract, heart and blood vessels, generative organs, and the nervous system. Only the respiratory organs were found to be very different from those in "perfect" animals, a fact recognized as such, particularly by Malpighi. The outstanding quality of their achievements is perhaps best illustrated through a comparison with other diligent students of insects.

The Swiss physician Johann von Muralt studied a range of insects in the course of several years.[42] The results of these studies were published in a lengthy series of entries, concerning individual species, in successive issues of the *Miscellanea curiosa* in 1682 and 1683 (see appendix 4). These descriptions, often very short, were assembled many years later into a book entitled *Dissertatio physica, de insectis, eorumque transmutatione,* published in 1718. Muralt studied and described a number of specimens minutely, examining their exteriors and such parts from their interiors as he could easily remove from their bodies. From the accompanying illustrations it is obvious that he occasionally applied some sort of magnifying instrument to his objects, albeit one of very small magnification. His dissections of insects were rather successful as far as the larger species were concerned, notably the field cricket and the mole cricket.[43] He provided a clear description and illustration of the latter's distinctive digestive tract, but it is in his treatment of the smaller species that the inadequacy of his skills became apparent. For such species as the louse he could only provide rather superficial descriptions of the exterior of his specimen.

Johann Franz Griendel von Ach used the attractive, and at the same time awe-inspiring, image of enlarged insects to promote the sale of his microscopes.[44] He was an optical instrument maker and in a booklet entitled *Micrographia nova* (1687) he introduced a new kind of microscope. The main feature of this instrument, as stated by Griendel, was an enlargement of the field of vision as compared to the fairly commonly available English type of microscope. In the preliminaries to his booklet Griendel reminded his readers first of the age-old dictum "nihil est in Intellectu, quod prius non fuerit in sensu";[45] he then described his newly developed microscope and ended with a discussion of a number of microscopic observations on several kinds of insects.

Griendel's microscopic observations cover a succession of insects for which he depicted only the outward appearance. Some specimens, such as the common fly, he treated rather elaborately, devoting separate chapters and engravings to the description and illustration not only of the general appearance, but also of specific details, that is, the head, proboscis, compound eye, and a leg. Like Hooke before him, Griendel also described several vegetable objects, among them the ligulate flowers of some composite plants and various seeds; textiles, hairs, and artificial objects. In fact, the majority of Griendel's illustrations are strongly reminiscent of Hooke's. Even though he may not have copied these without personal examination of similar specimens, it seems certain that Griendel relied heavily on contemporary publications, such as Hooke's *Micrographia,* for the preparation of his own booklet. One observation in this succession of familiar microscopic specimens for the layman betrays Griendel's reliance on current knowledge rather than on his own observations. It is a curious, discrepant report concerning the generation of the frog.[46] Griendel wrote that one day he had taken a drop of dew and placed it beneath a magnifying glass. He saw how this drop began to ferment and looking at the drop the next day he saw that a body and a head had already been formed. Two days later the body had acquired all the particulars of a frog. This absurd observation indicates Griendel's extreme gullibility, since he presumably concocted this observation on the basis of some obscure story.

Griendel's illustrations and descriptions are not unsatisfactory when taken at face value; when considered against the background of contemporary microscopic investigation, however, they are of inferior quality and offer no novel points of view, and no additional data of scientific interest. The examinations performed in England, the Low Countries, and Italy had by 1687 progressed far beyond the level of work that Griendel presented in his booklet, but since he wished particularly to stimulate the market for his merchandise, his effort was certainly adequate because a prospective buyer of a microscope became acquainted, in a pleasant way, with the possibilities of such an instrument.

Griendel's efforts were emulated some years later by Filippo Buonanni.[47] The latter's *Micrographia curiosa* was also largely devoted to the description of insects. Buonanni, a pupil of Athanasius Kircher and curator of Kircher's museum in Rome, enjoys a somewhat discreditable reputation in the history of science because he championed the cause of spontaneous generation on untenable grounds. His defense was based mainly on the erroneous view (even at the time that he advocated this opinion, contemporaries knew better) that mollusks have no heart and are bloodless. Buonanni derived his views to a large extent from classical, and a few contemporary, reports.

In his *Micrographia curiosa,* which appeared as an appendix to his *Observationes circa viventia* of 1691, Buonanni reminded the reader first of various passages in the works of classical writers such as Saint Augustine, with their allusions to the very small component parts of living beings. However, he wrote, although these very small parts had been impossible to study until recent times, a new world had now been laid bare with the help of the microscope.[48] Buonanni briefly introduced his readers to the works of contemporary explorers of that new world, discussing Hooke, Swammerdam, and Leeuwenhoek at some length, but barely mentioning his compatriot Malpighi. He also surveyed the development of the microscope and included an illustration and description of a microscope of his own design, for which he is probably best remembered.

Buonanni devoted the best part of his book to a lengthy discussion of the microscopic appearance of various subjects such as the mosquito. He had in fact, he wrote, made a personal study of a mosquito, but chose to illustrate his account with a copy of Swammerdam's representation of that insect, which had appeared in the latter's *Historia insectorum generalis,* the reason being that Swammerdam's illustration represented details that he had been unable to see.[49] The same applies presumably to the inclusion of copies of Hooke's illustrations of the louse and mite. Copying the illustrations from a colleague's book was certainly not an uncommon practice at the time, and Buonanni need therefore not be reproved. Historians of science, however, should be aware of such practices and should not bestow praise (as Buonanni was praised for the exceptional quality of his illustration of the mosquito)[50] where no praise is due.

Buonanni seems to have been particularly interested in the scales on the wings of butterflies, and he depicted numerous various forms of these. All in all, he gave a fair view of some of the subjects that had recently been investigated, such as the compound eye of the insects, but he relied so heavily on contemporary literature that his *Micrographia curiosa* is only slightly more than a review of the state of the art, and, moreover, a limited review as it is largely devoted to insects. However, he made personal examination of the subjects he discussed and enlarged on them to some extent. However, he chose to concentrate on aesthetically pleasing specimens and to avoid more discursive subjects.

Several other contemporary reviews of recent developments in microscopy reveal the authors' equally pronounced preference for the recent discovery of the details in the insect body. On the occasion of the publication of the first issue of the *Miscellanea curiosa* in 1670, Philip Jacob Sachs provided a review of microscopic science but enumerated only the microscopic achievements of Fontana, Borel, Kircher, and Hooke, from whose writings he selected mainly the descriptions of insects.[51]

Friedrich Schrader, on the other hand, surveyed the harvest of two decades of microscopic research more seriously in his *Dissertatio epistolica de microscopiorum usu in naturali scientia et anatome,* which was published in 1681. Before tackling his main subject, Schrader first discussed at some length the doctrine that all knowledge is derived from observation, both of the parts and structures of living beings and of the natural phenomena. Where the observation of the more delicate parts of the anatomy of living organisms was concerned, a great obstacle, and therefore the cause of many a mistake, had been the disproportion between the dimension of these parts and the senses. The microscope, wrote Schrader, had overcome this disproportion. Schrader discussed many instances of the amazingly intricate structure of living matter, taken from the works of Swammerdam, Leeuwenhoek, Borel, Hooke, Grew, and Malpighi. The majority of these concern the anatomy and appearance of insects, their innumerable eyes, the exquisite scales on their wings, and the delicacy of their anatomy. Justly, Schrader also discussed the criticisms launched at the microscope, such as Kerckring's objections, which have been discussed in relation to Malpighi's work (chapter 1). Schrader was obviously very much impressed with the recent discoveries and concluded enthusiastically "that no invention has been more brilliant and useful for anatomy than the invention of the microscope, whatever adversaries, unacquainted with these artifacts, say to the contrary."[52] However, amid a lengthy enumeration of microscopic achievements concerning the insect body he made only a cursory suggestion that knowledge of microscopic structure provides many a clue toward the explanation of function,[53] which, as we have discussed earlier, had been the main incentive for such investigators as Grew and Malpighi.

In conclusion it appears that most of the publications containing microscopic data on the natural history of insects were derivatives of earlier published works by Hooke and Swammerdam. Very few individuals attempted and achieved independent investigations. Among these were those of Giacinto Cestoni, which formed the basis for his joint publication with Giovan Cosimo Bonomo entitled *Osservazioni intorno a' pellicelli del corpo umano,* published in 1687, in which it was demonstrated that scabies is caused by a species of mite.

The part of the insect that attracted particularly wide attention was its compound eyes, which were not only studied by Hooke, Leeuwenhoek, and Swammerdam, but also by a number of men of lesser stature, among them Olaus Borch, Christian Mentzel, and Bidloo.[54] It is a curious fact that the very few contributions in the *Journal des sçavans* on microscopy were all concerned with the compound eyes of insects. Philippe de la Hire, the Abbé Catelan, and Louis Puget, as well as their fellow member of the Académie des Sciences François Poupart, investigated and speculated on the function of the compound eye.

It is a not too far-fetched speculation to assume that this research was prompted by académicien Perrault's insistence that insects lack all the faculties of sensation, except that of touch.[55] De la Hire took the initiative and demonstrated in 1678 that insects possess genuine organs of sight in the ocelli, single eyes situated at the anterior end of the head.[56] Two years later Catelan established microscopically that the two big bulges at the side of the insect head are composed of numerous elements similar to the ocelli, and are therefore compound eyes.[57] The elementary particulars of the structure of the compound eye were discovered and grasped by Catelan, as well as by Poupart and Puget, who published their findings some twenty years later. Notwithstanding the fact that the details of the mechanism of vision eluded the sight of these investigators, each discussed it at length. They agreed with Odierna and Hooke that only perpendicular rays may enter the separate eyes and so form an image. The question of whether the insect became aware of the thousands of slightly different images was a subject for happy speculation. Hooke had suggested that, although many thousands of images were formed on the retina, the insect only became aware of those things on which it concentrated its attention, through what Hooke called its "observing faculty."[58] This faculty might wander from facet to facet, he thought.

Poupart observed that in the dragonfly two branches of the tracheal tubes entered both compound eyes. This arrangement triggered Poupart's fantasy as to a possible mechanism with which the insect could adjust its eyes to the ever-changing demands of far and near sight. He wrote that the "libella could drive the Air contained in these canals into the eyes, to give a greater convexity to behold objects that are very near and on the contrary the Air is forced out of the Eyes again to flatten them when they look at remote objects." Poupart took some pains to test his idea and found that "this conjecture is not altogether frivolous, for having blown into the thick canals which are about the middle of the Body, the Eyes became considerably tumified, and by letting the air return, they became flat again."[59]

Puget, on the other hand, assumed that although every eye formed a separate image, the construction of the compound eye ensured that from the separate images one overall image was composed on the retina. He argued that the light "having traversed the cristallins, reunites in one single spot in the eye of the fly, to paint but one picture . . . and the convexity of the whole cornea is a necessary condition for causing this reunion, in that it directs to one point all the focal points of the cristallins."[60]

Antoni van Leeuwenhoek

Apart from a number of isolated excursions into problems of physics, crystallography, and geology,[61] Leeuwenhoek was primarily concerned with the intricate structure of living matter and—to a lesser extent—its bearing on the functioning of the body. He designed and manufactured his own microscopes, which were of a rather wayward design. In the course of his career he made hundreds of them, a practice peculiarly his own. A certain object was affixed to each microscope for observation (see figure 20) and was preserved for future reference. Indeed, by the time his microscopes were auctioned in 1747 many of them still carried specimens.[62] Leeuwenhoek published the particulars of the construction of his ordinary microscopes along with his description of two variant designs of his "aquatic" microscopes,[63] but never disclosed his methods for grinding his lenses. Over the years Leeuwenhoek became very skilled in making his microscopes and particularly in preparing the lenses and mounting them in their frames, which he deemed to be of the utmost importance.[64] As a result his later microscopes were of better quality than the earlier ones, and consequently Leeuwenhoek was able to improve on earlier observations, for instance, the structure of the lens in the vertebrate eye (see below).

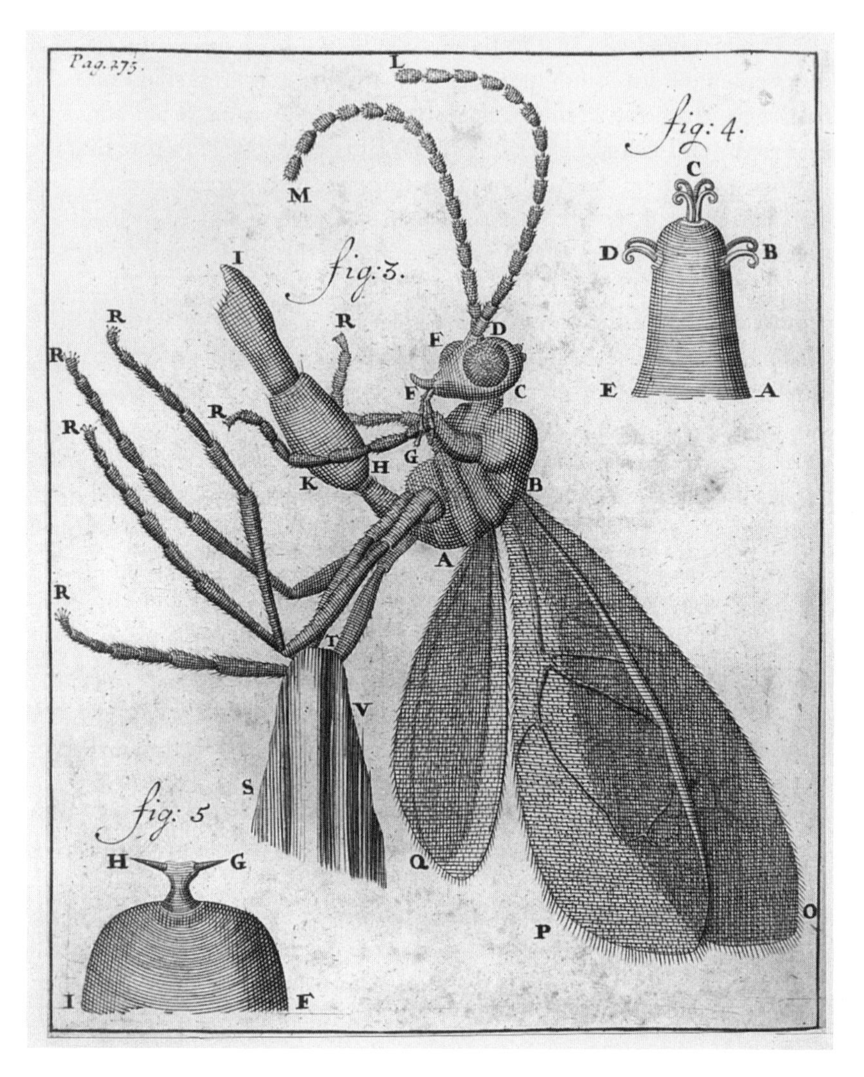

Fig. 20. An ichneumon fly attached to the point of the object holder of the microscope, illustration in Leeuwenhoek's letter of 26 October 1700, in his *Sevende vervolg der brieven* (1702).

Leeuwenhoek is universally attributed with having the best microscopes that were made prior to the nineteenth century.[65] Yet, judging from the extant specimens and those that were studied in the past for their optical properties, it is obvious that Leeuwenhoek owned a great many rather average microscopes,[66] and though it cannot be doubted that he also made and used some exceptionally good lenses, it is equally certain that other men, such as Swammerdam, Huygens, and Hooke, also produced and employed some very good lenses. Indeed, Swammerdam's observations on the particulars of the compound eye of insects, for instance, surpass those of Leeuwenhoek. These observations were made either with a microscope of his own fabrication or with a commercially acquired microscope. Huygens's descriptions and figures of microorganisms were certainly not inferior to those of Leeuwenhoek; rather, the reverse is the case, as will be discussed below. Huygens designed and constructed a new type of single-lens microscope, as discussed in chapter 1, on the strength of Leeuwenhoek's discovery of the spermatozoa. Hooke, as already indicated in chapter 3, successfully considered ways and means to perfect lenses for the simple microscope, prompted by the wish of the members of the Royal Society to verify for themselves Leeuwenhoek's remarkable observations. The result of such efforts was that, by the time Leeuwenhoek's career was well under way, microscopes of the same magnification and resolution as his own average instruments were readily available. Among them were the microscopes sold by the firms of Edmund Culpeper and the van Musschenbroek family.

Surveys of Leeuwenhoek's microscopic techniques indicate that he employed similar procedures to those of his fellow microscopists.[67] Usually he prized apart his objects (such as the muscle fibers), or cut slices of various materials, in a simple but effective way, but taking measures to ensure that the material could be cut. One method was drying the object before cutting, but in many cases he had to use all his ingenuity and skill in dissecting the specimens in some way, in order to discover its secrets. It is rather puzzling that some examples, such as the anatomy of the compound eyes in insects, appear to indicate that Leeuwenhoek's anatomical technique was rather inferior as compared with Swammerdam's, for example. Yet such a conclusion is contradicted by the fact that Leeuwenhoek was able to achieve the dissection of the cochineal insect, even detecting the particulars of the young ones contained within the egg, which in its turn was lodged within the parent body.[68] This was a feat requiring great dexterity and patient handling.

Leeuwenhoek often kept his preparations for a very long time, which caused them to desiccate, but he was well aware of the fact and described in minute detail the changes that occurred in such circumstances in muscle tissue.[69] In fact he even turned this liability into an asset, because he realized that in desiccated specimens some details were visible that were hard to distinguish in normal circumstances. The membranes surrounding the muscle fiber were thus demonstrated. If it were necessary for further investigation he restored, or so he thought, the preparation to its former state by putting a drop of water on it.

Leeuwenhoek was directed to his subjects in various ways. He owed the discovery of the spermatozoa, for instance, to a student who brought him the sperm of a sick man.[70] The primary discovery of the infusoria dates to the day that he crossed a nearby lake.[71] The full appreciation of this particular discovery, however, followed in the wake of an inquiry into the cause of the taste given by spices, at which time it appeared that the infusions he had made of pepper to study the taste-causing particles were infested with "animalcules."[72] Intrigued, he postponed the further investigation of taste and turned to the animalcules instead. Enquiries into the alleged death of leather-jackets (larvae of the daddy longlegs) on the advent of warm weather, and into signs of disease in fruit trees, led, respectively, to a detailed description of the life cycle of the daddy longlegs and the discovery of parthenogenesis.[73] Yet more often he was guided quite naturally from one subject to another, as when he studied the animalculae in tooth tartar, subsequently turning to various blemishes and excrecences on the human skin.[74] His unabated interest in the function of the spermatozoa prompted him to establish their presence in a wide range of animals and, in passing as it were, he discovered various novel details concerning their anatomy, as was the case with the cilia on the gills of the mussel.[75]

Leeuwenhoek's method of selecting subjects indicates that he had no preconsidered program of research, like Malpighi, and was not bent on documenting a specific hypothesis, like Swammerdam. He preferred to elucidate the details of nature's multifarious constructions, the "hidden invisibilities," as he often called them, and he approached that task in a truly Baconian fashion, making some lengthy and many shorter excursions into the three realms of nature. Whereas his choice of subjects was rather haphazard, his handling of individual subjects was consciously directed toward the elucidation of the purpose of a particular structure or object.

Generally speaking, Leeuwenhoek studied the microanatomy of nature, turning to the histology of the higher animals, examining brain, muscle, nerves, the texture of the lens of the eye, and so on. He also examined the details of insect anatomy and the anatomy of wood, taking great pains to satisfy himself as to the exact details. Much effort was put into the elucidation of the role of the "animalcules" in semen, in connection to which he also inspected a range of plant seeds. Curiously enough he studied the embryological development of animals or plants, a natural sequel to the discovery of the spermatozoa, barely at all.

It would be almost impossible to enumerate the subjects that Leeuwenhoek investigated. He studied nearly everything at hand, animate and inanimate, that caught his attention for one reason or another. A representative example of Leeuwenhoek's investigations is his work on the eye, both the compound eye of insects and the vertebrate eye.

Leeuwenhoek's studies concerning the organ of vision date back to the very beginning of his microscopic observations. Initially he studied the various tissues of the vertebrate eye microscopically and, except for the lens, was content to establish that globules predominated in each of these structures.[76] The lens appeared to him to be constructed from numerous very thin layers surrounding each other, and in each of these scales he discerned a circular pattern of lines. He likened this structure to a globe constructed from very thin paper layers, on each of which lines were drawn from pole to pole. The transparency of the lens was also discussed in relation to an earlier statement that tissues that appear white to the naked eye are in fact composed from transparent globules.[77] He suggested that the transparency of the lens resulted from the neat arrangement of the globules, so that the light could pass through it unhindered.

Ten years later the subject of the structure of the lens was reinvestigated, because Leeuwenhoek believed that he had not yet completely grasped it. He devoted one long letter to this topic and in a few subsequent letters made some additional remarks.[78] In the course of his renewed research he discovered a thin membrane surrounding the lens, and he decided that this membrane must be composed of very small vessels or threads, a conclusion formed by analogy with other membranes,[79] but which he was unable to confirm through observation. His earlier conception of the construction of the lens was confirmed, and he was able to specify some details. He discerned a clear pattern in the distribution of the circles in each separate layer of the lens, which was accurately described and depicted: a three-poled pattern in the cow, pig, and some other animals, and a two-poled pattern in hare and rabbit. He could not definitely determine the pattern in birds and fishes or in human lenses. Leeuwenhoek imagined that each scale was formed from one single "thread" intricately wound to a definite pattern. To illustrate this idea he constructed a model by putting three pins into a hand ball, which figured as the poles, and glued a woolen thread around its surface in the pattern he had described.

At the time of his earlier investigations Leeuwenhoek had thought that the lens and the other parts of the eye were composed of globules. Since he had revised his ideas concerning the units of construction for living matter in the intervening years, as a result of more exacting examinations (see below), he now explained the ever-present globules as the disrupted parts of threads, and he speculated whether the threads might not be composed of still smaller threads, very much in the same way as the muscle fibers (see chapter 4).

At the same time Leeuwenhoek also reinvestigated the structure of the cornea; his earlier ideas were confirmed, but his description differed from the former one in that he described the cornea as composed of numerous immensely fine fibers, densely packed rather than composed of globules. Another brief study of the cornea some eight years later brought about another slight change in his conception of its structure.[80] He now perceived it as being composed of hundreds of thin films stacked one on the other.

Leeuwenhoek's final inquiries concerning the vertebrate eye occurred more than twenty years later.[81] While studying some parts of a whale he also investigated the eyes, and was much struck by the solid construction of its lens. As was his custom, he endeavored to explain the reason for this condition, in this case with a reference to the whale's way of life. When a whale dives deep, enormous pressure is exercised on the eyes, which he calculated at 23,100 pounds (about 11,000 kg).[82]

Leeuwenhoek had made a short reference to an investigation of the compound eye of the bee in his very first letter to the Royal Society, and therein put forward the rather quaint theory that the hexagonal shape of the honeycomb is the result of the fact that bees perceive the world around them cut up into hexagons because of the shape of their eyes,[83] a view that was later ridiculed by Swammerdam.[84] Leeuwenhoek did not investigate the compound eye for another twenty years, but when in the summer of 1693 he had occasion to investigate some particulars of the dragonfly, he also studied its eyes. In the course of that study he put the cleansed outer, horny layer (the cornea) before his microscope and at the same time, he "placed a burning candle a little away from me, in such a way that the light of a candle had to pass through the Cornea, and thus represented the flame of the candle upside down through the Cornea, so that I saw not one, but several hundreds of candle flames; nay, I even saw them so plainly, however small they were, that could distinguish in each of them the movement of the candle flame."[85]

Leeuwenhoek was much affected by the perfection of the multifaceted cornea and initially concentrated his research on the optical properties of the cornea and the number of facets in various species, a subject to which he returned several times in later years.[86] It was only later that he attempted to find out the exact structure of the substance adjacent to the cornea, which he had at first only described as being of a stringy nature. Like Odierna and Swammerdam, he finally resolved this substance into numerous particles of almost equal length, each one slightly broader at one end than at the other. He decided "that each of this large number of particles which I observed was an optic nerve and that the thick and round end of the optic nerves had been placed in the small cavity in the Cornea, briefly: there are as many optic nerves as there are facets in the Cornea."[87]

On the outside of these elements he discerned small stripes, or "wrinkles," which reminded him of the structure of muscles (see figure 21). For this reason he thought that the "nerves" in the compound eye could also be stretched or contracted as occasion demanded. Leeuwenhoek described one novel feature. In the course of his investigation into the eye of a shrimp, he observed between the "nerves" a great number of clear particles, which he described as long and bent. He decided that these were the "cristalline humor," that is, the lens, of each separate eye within the compound. As their form was not perfectly round at the time of inspection he supposed that they had become slightly disfigured as a result of his dissection of the eye. In their original round form they would have fitted perfectly into the inward curves of the cornea.[88]

Whereas Leeuwenhoek described in minute detail the various structures in both the vertebrate and insect eyes, he advanced no theories as to the optical effects within the eye or the mechanism of vision. The contours of the characteristics of Leeuwenhoek's method already appear in this survey of his observations and speculations concerning the visual organs: that is, frequent returns to the same subject, reasoning by analogy, comparison with physical models, and quantitative analysis. The emphasis in his work appears to be on the lucid description of structure, backed by a definite view on the organization of living matter (which we have examined in chapter 2), and the correlation between the construction and the qualities of specific structures.

Leeuwenhoek's initial motive for turning to the microscopic investigation of the world about him will most likely remain forever obscure. Although it has been suggested by English writers that Leeuwenhoek's visit to England in 1667, at which time Hooke's *Micrographia* was creating a

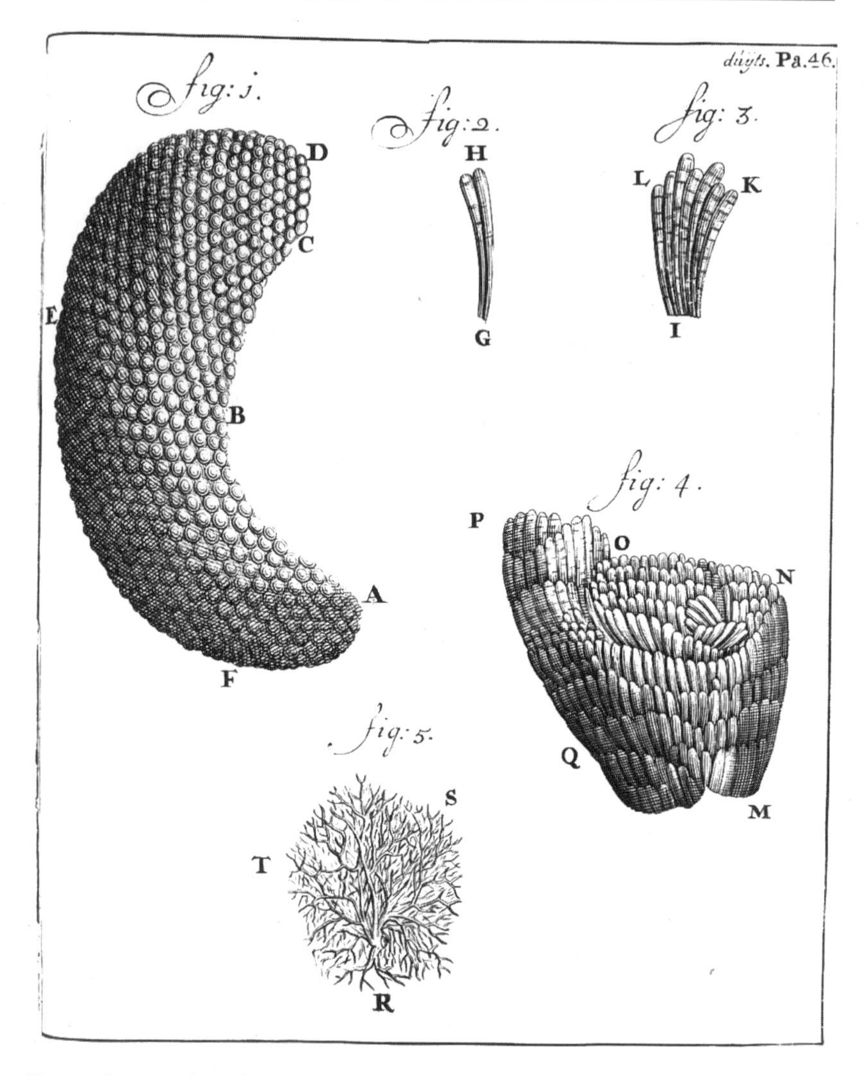

Fig. 21. Leeuwenhoek's representation of the cornea and omatidia of the compound eye of the beetle, in his letter of 9 May 1698, in his *Sevende vervolg der brieven* (1702).

furor there, may have been seminal in this respect,[89] such a theory remains conjectural. Yet it is obvious that the enthusiastic reception of his observations by the community of the learned stimulated him to concentrate totally on the subject. Moreover, he never tired of his researches because of the real pleasure it afforded him to look at nature's intricate handiwork.

It is therefore not surprising that such a man, armed with insatiable curiosity, a good microscope, and tremendous perseverance, should discover multitudinous structures and a host of organisms that others had not seen before him and still had difficulty in discerning once he had drawn their attention to their existence. Leeuwenhoek was unquestionably gratified by the attention of his contemporaries and the interest of scientists in his work, although he encountered a good deal of opposition. Like Hooke before him, Leeuwenhoek was rather disappointed by the fact that microscopic research was no longer very popular at the end of his life. In response to an enquiry from Leibniz about why he did not found a school, he replied that in Leiden students had become enthusiastic proponents of the microscope, and had taken lessons in lens grinding, but that in the end nothing had come of it, because no money was to made from this activity.[90]

Leeuwenhoek explored nature in the Baconian style, to which he was characteristically partial, a partiality that was certainly strengthened by his lasting contact with the Royal Society. There is no reason to think that he had launched his inquiries with a specific purpose in mind. In the course of his career, however, amid numerous sidelines he championed a few topics, especially the spermatozoa, spontaneous generation, the capillaries of the blood, and microorganisms. The discovery of specific entities such as the spermatozoa encouraged a program of research to settle their role in generation, as well as a program to refute spontaneous generation. His research therefore created its own range, the more so as he was not content until he had confirmed, through personal investigation, that a particular species did not constitute an exception to the rules he had decided on. As a result he continued with his routine to the point of tedious repetition. Incidently as it were, he discovered a great many particulars concerning the microscopic anatomy of numerous organisms. This fact, though, was scarcely recognized until Leeuwenhoek's writings were clarified with the help of modern biological insight and terminology.

Leeuwenhoek was undoubtedly conversant with the gist of the Cartesian conception of nature, but he only applied these notions to a limited number of topics: the corpuscular nature of organic matter, the cause of taste, the transmission of nerve stimuli, growth by means of superposition of particles, and the mechanism of transportation of fluids through plant vessels. It is striking that Leeuwenhoek proposed and discussed these mechanisms in some detail chiefly in the course of his early research (i.e., during the seventies and early eighties). In subsequent years he alluded to these mechanisms without further comment (the cause of taste, the mechanism of growth) or they disappeared from his writings altogether. The latter happened to the mechanism of transmission of nerve stimuli. Whereas in the 1670s he had envisaged transmission through the collision of particles within the lumen of the nerves, in 1717 he wrote about the nerve fluid flowing through the lumina in the nerve fibers.[91]

Wherever he looked, Leeuwenhoek observed organic matter to be minutely structured. In the cases in which no distinctive structures could be perceived, as in the spermatozoa and the microorganisms, the minute cilia on their exterior and the globules in their interior betokened structure, if the fact of their industrious movements had not already been sufficiently convincing. Indeed, he wrote,

> Should we not stand amazed by the incomprehensible multitude, and smallness of parts of which such a tail [of the spermatozoon] consists, and that chiefly when we establish, that such a thin tail must be fitted with so many elements in proportion to the tail of big animals, for the tail to move adroitly, and in all directions, and that each of their little parts, must not only have its particular muscles, but also nerves, and veins, which carry the nourishment, in short, the parts, and their smallness, of which the bodies are composed, are unthinkable to us.[92]

The probability of structures of ever-diminishing size was supported by a number of observations, for instance, on the trachea in insects, which seemed to disappear from his sight as he traced their diminutive ramifications. On the other hand a certain constancy of scale seemed to prevail in nature. In numerous different animals Leeuwenhoek observed the spermatozoa, and found their shape and form to be rather uniform. The size of the spermatozoa bore no relation at all to the size of the adult animal.[93]

The same holds good for the size of the blood corpuscles in various animals and the diameter of the muscle fibers.[94] Nevertheless Leeuwenhoek was absolutely convinced that the structures of organic matter extended into infinity. It might seem a convenient loophole in disputes, such as that with Garden discussed in chapter 2, to counter objections with the argument that the details concerned were just too small to be seen with the microscope, but to Leeuwenhoek it was an undoubtable fact. Consequently, the mechanisms of nature operated on a level beyond that which was microscopically visible. Such mechanisms therefore were, and remained, elusive. That being so, Leeuwenhoek's concern definitely shifted from the operative to the structural.

Leeuwenhoek was supremely concerned with structure and with the "consequences les plus immédiates,"[95] that is, with the purpose of a certain structure. His ideas as to the function of parts and structures usually took shape soon after, or even during, the first encounter with a given subject and, once formed, were not fundamentally changed over the years. This supports the idea that Leeuwenhoek was ultimately more interested in the discovery and exact description of structure than in the elucidation of the fundamental mechanisms of organic nature.

Microorganisms

Antoni van Leeuwenhoek made what turned out to be his most momentous discovery in the summer of 1674. In a drop of water, taken from one of the lakes near his home, he observed long green threads and numerous "little animals" flitting about. These animalcules were in his estimation at least a thousand times smaller than any of the small creatures he had up to that moment observed through his microscope. Although Leeuwenhoek was surprised by his discovery of such minute organisms, and came across similar organisms in other samples in the following years,[96] his interest was not aroused until he accidentally discovered, in April 1676, that the animalcules appear in unbelievable numbers in infusions of a great variety of plant materials. This discovery initiated a period of intensive research summarized and duly published in his letter of 9 October 1676. For posterity, the discovery of the infusoria and microorganisms in general really dates from the publication of this letter. In it Leeuwenhoek described a number of unicellular organisms observed in various environments ranging from natural habitats such as rainwater and water from ditches and canals, to artificially created environments such as infusions of pepper and ginger. For Leeuwenhoek the sight of the numerous little animals in a drop of water was "among all the marvels that I have discovered in nature, the most marvellous of all."[97]

Among the recipients of Leeuwenhoek's letter the discovery of the microorganisms caused great excitement and incredulity as well. As already mentioned in chapter 3, the members of the Royal Society requested Nehemiah Grew to prepare a demonstration of the animalcules, on the arrival of Leeuwenhoek's letter in April 1677. Grew, however, was unable to do so, either because the magnification provided by his microscope lenses was too low, or because the infusions were not allowed enough time to develop. Some months later the request was deputed to Hooke, who developed a new simple microscope, after the design by Leeuwenhoek, especially for the occasion.[98] The desired demonstration was thereupon achieved, and the members of the Royal Society could observe the animalcules for themselves.[99] It is reported that even the king of England made a request to be shown these unbelievably small living beings.[100] However, although the virtuosi were enthusiastic and discussed the discovery for some time,[101] none of the members began original investigations at that period. Certainly, Hooke prepared some infusions and studied them superficially,[102] but these only served as a demonstration and affirmation of Leeuwenhoek's research rather than as an original enquiry, like his earlier endeavors. Even so, the subject of the animalcules was eagerly taken up by contemporary playwrights, who contrasted the insignificance of these creatures to the pretentiousness of the virtuosi, to raise a good laugh from the audience attending their satirical plays.[103]

The news of the discovery of the infusoria also provoked a good deal of interest in other parts of learned Europe. Constantijn Huygens the elder, who had been a patron of Leeuwenhoek,[104] asked his son Christiaan to translate a letter by Leeuwenhoek into French in order to communicate it to the Académie des Sciences. This request had the perhaps unexpected result of rousing Christiaan's and his brother Constantijn's interest in infusoria, and they also began to study them.[105] Although this particular letter of Leeuwenhoek's did not reach the Académie, Christiaan demonstrated the infusoria to that body in August of 1678.[106] Huygens's immediate interest in the microorganisms was shared by a number of men within the Parisian scientific elite, such as Ole Rømer, Jean Picard, and Philippe de la Hire.[107] While his contemporaries' interest quickly evaporated, Huygens accomplished serious investigations on these bodies. Indeed, Nicolaas Hartsoeker wrote to Huygens in September 1679, only just over a year after Huygens had first demonstrated the infusoria, "I do not doubt that the French curiosity concerning the microscopes has disappeared entirely. As far as I am concerned, I begin to understand, I think, that one would have to peer through it for a long time before becoming any the wiser."[108]

The only investigator who published on the microorganisms prior to the 1690s was Leeuwenhoek himself.[109] From 1693 onwards several investigators contributed to the knowledge concerning the infusoria. Their papers, few and far between, were published in the *Philosophical Transactions*[110] and the *Memoires d'Académie des Sciences*,[111] but these minor works were soon surpassed by Louis Joblot's monograph on the infusoria, entitled *Descriptions et usages de plusieurs nouveaux microscopes . . . avec de nouvelles observations,* which was published in 1718.[112] The relative surge of interest in the infusorians within the circle of the Royal Society may have been created by two lectures, which Hooke delivered in 1692 and 1693, respectively, to the assembled members.[113] As already discussed in the introduction to the present enquiry, he argued in both lectures that microscopic research was of immense value for the elucidation of the "internall mechanism of Bodys," and lamented the fact that hardly anyone currently undertook such investigations.[114]

Although Leeuwenhoek's discovery of the infusoria met with immediate response throughout Europe, the discovery was not wholly unexpected within scientific circles. The infinite divisibility of matter, the infinity of the cosmos, the continuity of the Chain of Being, and related subjects had been widely discussed before 9 October 1676. Indeed, three years before, Nicolas Malebranche had surmised, in an often quoted passage, that the size of animals, like that of matter, might be infinitely small.[115] In other words, the existence of such minute organisms had already been accepted as a theoretical possibility before they were actually observed.

Let us turn to the main topics in the investigation of infusoria as they appear from the investigations published by several men and from Huygens's manuscript notes on this subject.[116] Although Huygens studied several kinds of infusions simultaneously for a prolonged period from 1678 up to 1679, and again for several months in 1692, he did not assemble his jottings into a coherent picture. However, a survey of his notes is relevant, since his studies touched on several of the interesting features of these organisms that occupied the minds of many other scientists.

Leeuwenhoek, Huygens, Joblot, and their English fellow investigators usually took a few drops from one of their infusions and studied them microscopically, noting the number of animalcules, the various kinds, and any other interesting particulars. Leeuwenhoek, Joblot, and Huygens—presumably the other investigators did the same—repeated the sampling over a period of time. By going about their studies in this way they observed that in such infusions the numbers of animalcules depended on the length of time passed after the infusion was made, the amount of water that had been evaporated, the temperature of the water, and so on. However, such observations are more or less drowned in the mass of detailed descriptions of the appearance of the individual kinds of organisms. Leeuwenhoek discovered and described a large assortment of various kinds of infusoria, from diverse environments, of a variety of forms, and of different living conditions. By hindsight it has also been concluded that Leeuwenhoek discovered bacteria.[117] However, neither Leeuwenhoek nor his contemporaries distinguished these much smaller animalcules from the bigger kinds. On the contrary, Leeuwenhoek often thought that these smaller "dierkens" might be the young of the bigger ones.[118] Although Leeuwenhoek discriminated the various animalcules by drawing attention to their different characteristics, he did not classify the various kinds by assigning them proper names. Although Leeuwenhoek published some drawings of the infusoria, *Volvox sp.* and bryozoa among others, most of the creatures were only described.

Huygens, on the other hand, made lucid sketches of his specimens (see figure 22), ascribed a letter code to each drawing, and referred to them by that letter in subsequent observations. A representative entry in his note book thus reads "Old water of pepper, many H, several N, 2 or 3 M with a tail, a great number of small ones like black points, which move a lot, like yesterday, a few K."[119]

Joblot combined the two strategies and published both fairly accurate illustrations of the specimens he had observed and descriptions. The most noticeable and novel feature of his treatise was no doubt the fact that he attempted to classify the various kinds of organisms. He named the animalcules after objects and animals to which, in his opinion, they showed a certain resemblance, such as "the swan," "the silver bagpipe," "the glutton," "the slipper," and so on. Although he was not very accurate in his identifications, sometimes applying the same name to different organisms, his efforts at nomenclature were important in that certain kinds thus acquired an identity that distinguished them from the mass of infusoria. In other respects Joblot's observations ran on in much the same vein as those of Leeuwenhoek, paying particular attention to the movements of the animalcules.

By diligently working through Leeuwenhoek's letters and identifying each organism (as Dobell did, and the editors of Leeuwenhoek's collected letters still do) it has become apparent that Leeuwenhoek saw a large number of different species. Since he described them with great accuracy many of them can now be identified. This fact has led to an overappreciation of Leeuwenhoek's work on the part of some historians of science.[120] Identifying the range of Leeuwenhoek's researches with present knowledge, it is easily overlooked that his contemporaries were in fact presented with a mass of details, difficult to interpret. Just because Leeuwenhoek kept referring to his animalcules in terms like "the pear-shaped ones," the very extent of his researches was obscured for his contemporaries. It is only Joblot's treatise, in which he depicted most of the organisms he saw and named the majority of these, that permits the fact that the infusoria form

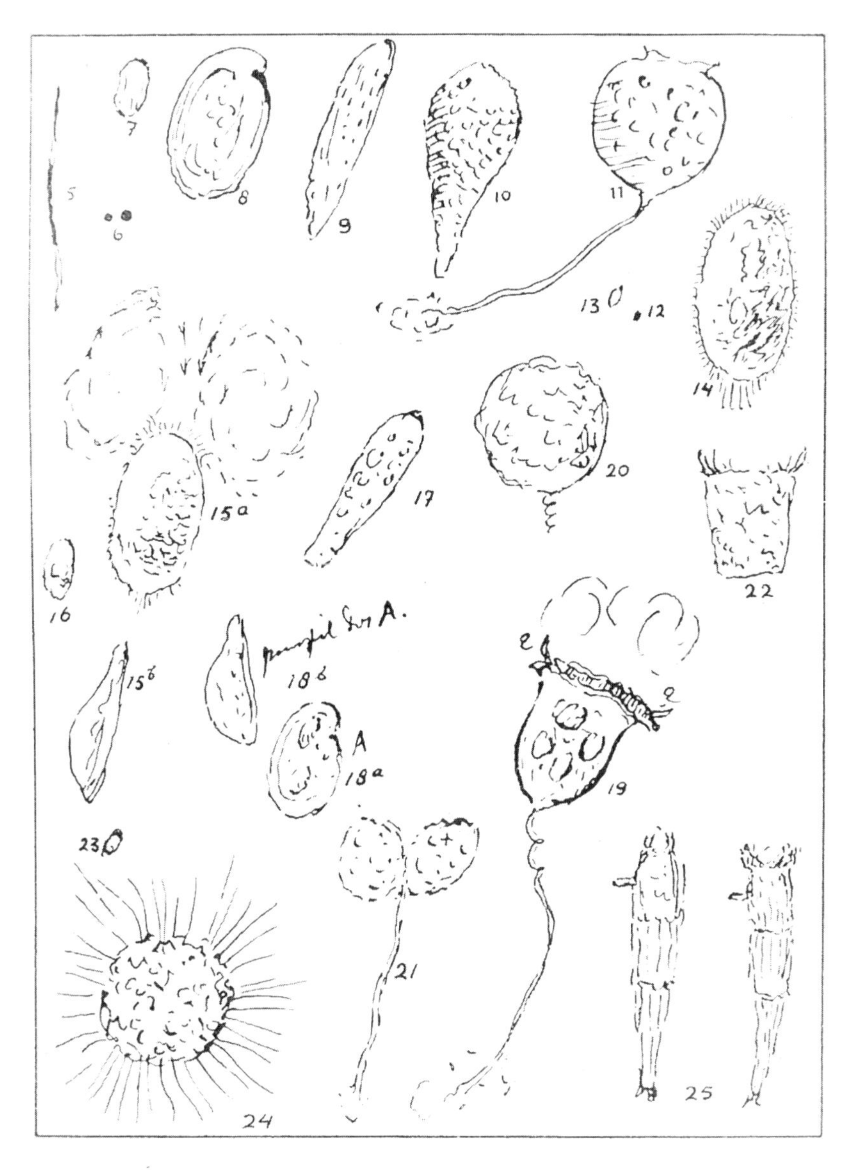

Fig. 22. A compilation of some of Huygens's drawings of microorganisms to be found in his manuscripts and letters (Museum Boerhaave, Leiden).

a large group of organisms (consisting of numerous different kinds) suddenly to stand out.

The various investigators also made notes that demonstrate their recognition that new kinds of animalcules appear, and others disappear from the infusions as time passes. What is not so clear, however, is whether these observations made any impression in the sense that they realized that a natural phenomenon was being observed. Huygens only remarked once on the fact that he saw various kinds of animalcules appear in succession. He noted, on inspection of his own pepper infusion and the one he had been given by Rømer, "Thus the same animals have got in this water, as in that of Mr. Rømer of the same pepper, but much later in mine because this was put much later to infuse."[121]

Leeuwenhoek and Joblot did not remark on the succession of different kinds of infusoria, but an unknown Englishman did. This man, who wanted to remain anonymous and has succeeded in remaining so despite the efforts of at least one historian of science to discover his identity,[122] disclosed that he had noted "a constant order of their appearance,"[123] but did not enlarge on this phenomenon, only using it as an argument in favor of his notion on the origin of the infusoria.

The descriptions of the microscopists concerning the appearance of the infusoria and other kinds of microorganisms concentrated, naturally, on their general form (round, pear-shaped, funnel-like, etc.) and their way of moving. However, some details of their structure were also observed, for instance, the globular particles from which they seemed to be composed, the pulsating vacuole, and particularly the cilia covering the outside of many specimens. Although the nature of these particulars was not clear (they were often denoted as "little feet" and attributed with a minute internal structure) their function was partly understood. Huygens noticed how some of the creatures propelled minute particles of food into their mouths by means of a water current created by the cilia. Indeed, he wrote, "they sometimes have before them a kind of circulation of particles and dirt."[124] Leeuwenhoek reached a similar conclusion: he saw in some specimens rows of these "legs" around their mouths, and estimated that the cilia brought about a steady stream of water toward the organisms so that food particles would come into their reach.[125]

Huygens also noted that the "little feet" moved continuously whereas the animal did not advance forward or backwards. Therefore, when considering how they transported themselves from one place to another, Huygens rejected the idea that the cilia could have anything to do with movement. He could not imagine anything else "than that they have the ability to inflate a part of their body and that successively from head to tail, so that this swelling or bump travels very rapidly the length of their bodies and then starts again at the head and this suffices to move the animal forward."[126] This was a novel idea, which remained buried in Huygens's manuscripts. For an explanation of movement in infusoria, his contemporaries relied on an analogy with familiar animals, which suggested that infusorians move in much the same way as do other animals, either with feet or fins, for which function the cilia seemed the most likely candidates.

Apart from studying the infusoria in ideal natural circumstances, their reactions to hazardous circumstances were also studied. Leeuwenhoek noted that as the liquid that contained the animalcules evaporated, "they burst asunder whenever the water happened to run off them." Any time this happened Leeuwenhoek saw that "the globules and a watery fluid flowed away on all sides."[127] This phenomenon was also noted by other microscopists, such as Joblot.[128] Even though most men inferred from the swift and characteristic movements of the animalculae that these were indeed animals, Edmund King felt that any such conclusion had to be substantiated. He performed some experiments to prove the point.[129] He subjected several kinds of animalcules to extreme conditions, by adding drops of vitriol, salt solution, ink, blood, urine, and sugar solution to the samples of his infusions under the microscope, and observed their reactions as they came into contact with these substances. Initially the reactions consisted of wild movements; sometimes the organism shrank (as in the case of salt solutions), and as a rule the animalcules died rather quickly. These experiments were repeated by the anonymous Englishman who observed the same reactions as King. Both concluded from the reactions observed that the infusoria were indeed animals.

Huygens also experimented with the effect of temperature on these organisms: in the winter of 1678 a phial was left to freeze on the windowsill of his rooms, while another was brought to the boil, so that in both infusions all the animalcules were killed. He found that two days later the former phial once more contained animalcules, whereas in the latter no living beings could be found after three days.[130]

Experiments like those discussed above were associated with the generation of the animalcules. The origin and reproduction of the animalcules was a much discussed issue, which presented a baffling problem. Current theories claimed that the animalcules came into the water from the air. It was presumed that either the organisms floated through the air and were attracted by the smell of the infusion, or that the animalcules were in fact the larvae of a variety of minute flies deposited in the water, which in their turn took to the air in due course.

Christiaan Huygens was an advocate of the former theory. The most convincing argument in his view was that the same kinds of animalcules emerge in different kinds of infusions. He tried to substantiate this proposition by comparing the development of animalcules in closed and open phials. He prepared a pepper infusion and put part of it in an open phial and another part in a closed one, to see if these would bring forth the animalcula. He noted that many eellike organisms (i.e., bacteria) and other creatures had appeared in the open phial after two days, but that these could also be discerned in the closed phial, albeit in smaller numbers. A week later Huygens noted that quite a number of different organisms were to be observed in both phials. In spite of these results Huygens maintained that the animalcules descend into the water from outside or are deposited there as the eggs of some kinds of insects.[131]

He repeated this experiment thirteen years later. This time he noted that after only two days living creatures could be observed in the open phial, while it was over three weeks before any organisms could be discerned in the closed phial. Huygens was loath to relinquish his theory and concluded that these organisms, or their seeds, must have entered the phial through pores in the leather that he had used to seal the infusion from the air.[132]

Leeuwenhoek performed very similar experiments in June 1680, with identical results. These experiments were designed not so much with the question of the origin of the animalcules in mind, but rather as a verification of Redi's experiments demonstrating that meat or liquid contained in a closed bottle does not generate living creatures. Leeuwenhoek, who contrary to expectation did find animalcules in the stoppered phial, explained that Redi's experiments were concerned only with worms and maggots, whereas he was concerned with microscopic organisms, but he did not elaborate on his own results.[133]

In France similar experiments were reported. One of the philosophes related his results in the *Histoire et memoires de l'Académie Royale* of 1707. These experiments are variants of the experiments subsequently reported by Joblot. The philosophe mixed water and manure and then boiled it for some time. When the mixture had somewhat cooled, he poured it into two phials. Into one of these he also put a few drops of water containing animalcules and then he plugged both phials. Eight days later he observed that the water in the phial to which living animalcules had been added now teemed with them, while the water in the other phial contained none. This experimenter concluded from these experiments that the animalcules do not come from the eggs of airborne animals, but multiply in the infusion. Indeed, he had also observed animalcules coupled in pairs in his infusions. Apparently some members of the Académie had offered the suggestion that these animalcules were engaged in battle. The author of the report objected, "But do they never fight in any other fashion than two by two?"[134]

A decade later Joblot published his monograph on the infusoria, in which he reported his experiments concerning the origin of the animalcules. He prepared two infusions from hay, one sealed, the other left open to the air. He found, when comparing these after some time had elapsed, that both infusions yielded equal numbers of animalcules. He concluded that "eggs" from which the animalcules issued were present on the hay stems. His subsequent experiments departed from this point. He boiled a quantity of hay for some time and split the resulting infusion into two parts. One he sealed from the air while still warm, the other was left open. He found that animalcules soon began to appear in the open infusion, while in the sealed infusion no organisms could be detected. Thus, boiling had killed the "eggs" present on the hay, but animalcules could still enter the infusion from the air and live there, because when Joblot opened the sealed infusion animalcules soon began to appear. The conclusion at which Joblot now arrived was that the animalcules originated from "eggs" floating in the air. The air, in Joblot's opinion, was full of minute organisms that settle on plants or drop their eggs onto plants from the air. These eggs constitute the beginning of the infusoria.

The various experiments concerning the generation of the infusoria between 1680 and 1720 were much influenced by the almost universal rejection of spontaneous generation. The rejection of that theory meant that the infusoria could not have arisen de novo in the infusions. At the same time, as a consequence of the research methods employed (i.e., the fairly brief inspection of a small sample), the methods of reproduction in infusoria now recognized were only occasionally observed. Even more confusing, all the organisms of one kind seemed to be of one size; Huygens noted that he never saw these animals grow large from small, at best some were half the size of others.[135] Therefore, the best solution to the problem of the infusoria's origin appeared to be an agent outside the infusion. In the face of the accumulated facts, observations, and experiments, the generally accepted conclusion concerning the origin of the animalcules in infusions was that "most of them are the product of the Spawn of some Volatile Parents, and generated like Gnats, and many other sorts of Flies, which are bred, and undergo many changes in the Water, before they take wing."[136] The unknown author of this statement yet felt that this might not be the ultimate or only answer to the question and added that these organisms were so small that they, or their spawn, might be lifted from one pool with the evaporating water, to be blown away with the wind and to be dropped with the rain in another pool.[137]

Obviously, the principal weakness of the invisible insect theory was that no trace of such airborne animals could be found. This was felt very keenly by many investigators, among them Huygens,[138] but as no other solution (save for spontaneous generation) presented itself, it was accepted for the time being. Nevertheless, data bearing on the multiplication of animalcules were incidently noted. Leeuwenhoek, Huygens, the English and French anonymous, as well as Joblot all observed the conjunction of individuals of the same kind, which they interpreted as the first stage of sexual reproduction. Leeuwenhoek observed how daughter colonies took shape within a parent colony of *Volvox sp.* and also noticed that some organisms could protect themselves against desiccation by means of a firm coating around their bodies.

Although Leeuwenhoek undoubtedly contributed the major part of the publications concerning the infusoria during the period covered in the present enquiry, the quality of concurrent research was certainly equal to his. Even so, the number of scholars actively engaged in research was very small in this subject too. Although this small body of men gathered among them quite a lot of data on the infusoria, neither they—nor anyone else at that time—realized that these organisms constituted a wholly different group of living beings.

The scope of seventeenth- (and early-eighteenth-) century research extended to the description and portrayal of numerous forms, the discovery of their various habitats and experiments designed to discover their origin. About the structure of their bodies only one thing was certain, that many kinds were beset with very thin extensions of their surface. Along with Leeuwenhoek many persons assumed, for reasons of analogy, that the inner structure of the animalcules consisted of very much the same organs as those of larger animal bodies. The method of research employed by the early investigators, that is, preparing an infusion, taking a small sample from it, and observing whatever seemed interesting about the specimens moving about in front of the microscope lens, could in fact hardly lead to any other results. With such a method there was only a very small chance of observing reproduction by division or the process of sexual generation. Consequently an unsatisfactory mode of generation was, somewhat grudgingly, postulated by the investigators involved in this subject.

In conclusion, the impact of the discovery of microorganisms on contemporary science was rather mixed. On the one hand, their very existence was welcomed as proof of the principle of plenitude and the continuity of the Great Chain of Being. The enormous diversity in habitus, and therefore the large number of different species, was duly noted, but not systematically exploited, until well into the eighteenth century. On the other hand, after the initial amazement at their minuteness, variety of form, and ceaseless movement had passed, the animalcules appeared rather dull since nothing much could be learned from observing them in the ways then employed. Moreover, the animalcules proved an embarrassment to generation theories since they could only be fitted into the general scheme of things in a very inelegant way.

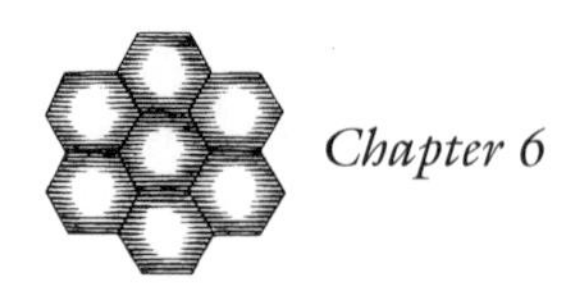

Chapter 6

Five Heroes of Microscopic Science

HAVING EXPLORED the scope of microscopic science in the seventeenth century in considerable detail, we should now return to the origins of this enquiry, that is, to the difference between Robert Hooke's two statements quoted in the Introduction. In the first, dating from 1664, Hooke foresaw that the microscope would offer splendid opportunities for scientific investigation, especially in the life sciences, while less than thirty years later, in 1691, he complained that the microscope was used mainly for pastime and diversion rather than for serious scientific research. The main objective of the foregoing analysis was to find an answer to the question why it was that microscopy flourished solely in the intervening decades. During this period five men towered among their peers; thus it is appropriate to appraise their work separately.

Comparing the substance of the microscopic output of these five, it is obvious that the results of their investigations were rather similar, although their research differed in scope. Hooke's investigations in particular are, by comparison, rather meager, as he investigated only a limited number of objects and only their outward appearance. The other four microscopists penetrated much further into the structure of living beings. The microscopic anatomy of plants was explored in great detail by Malpighi, Grew, and Leeuwenhoek with equal expertise. Malpighi, Leeuwenhoek, and Swammerdam each studied the anatomy of insects, Swammerdam's investigations, especially, were exhaustive. Malpighi and Leeuwenhoek both investigated various subjects related to the fabric of human and animal organs.

From the foregoing analyses of the works by these five scientists it has also emerged that each interpreted his results in a different context. Hooke presented his microscopic observations as a demonstration of the advantages in the employment of instruments in scientific investigation, namely, that with the help of such aids phenomena invisible to man's senses could be recorded or made visible. Indeed, the magnificent descriptions and equally impressive illustrations in *Micrographia* amply demonstrated his point. Although he discussed the function of various structures in relation to the physiology of the organisms, Hooke used his data to substantiate the corpuscular hypothesis of nature by relating individual observations to a specific level of organization of mineral or organic matter. In *Micrographia* Hooke clearly itemized his argument up to the level of what he regarded as the simplest form of life—mold—but from that level upwards the picture he presented was only fragmentary.

Grew concentrated his microscopic investigations exclusively on plants, but complemented his ocular examinations with chemical analyses. Among the microscopists Grew's dual approach was certainly unique. In his magnum opus, *The anatomy of plants* (1682), Grew attempted to merge his knowledge of plant structure and the results of his chemical analyses with the principles of the corpuscular and mechanical philosophies in such a way that the formation of organic matter from inorganic, the growth of organic matter into vegetable structures, and the operation of these structures in the service of the life processes of the plant formed an integrated whole.

Both Grew's and Malpighi's investigations of plants were aimed at the elucidation of their life processes. Unlike those of Grew, Malpighi's investigations formed part of a much grander undertaking: the explanation of the operation of all organic matter. In that respect Malpighi's scope paralleled Hooke's, but in contrast to Hooke he did not reduce his observations to the level of the basic constituents of organic matter—the corpuscles—but only attempted to explain the operation of organic matter on the basis of its microscopically visible characteristics. He realized in the course of his career that his premise—that by examining the microscopic structure of organs he would discover how they operated—was too optimistic, because he found that he could not resolve the finest details of his specimens. Nonetheless, his original notion, that the processes of life were brought about by the mechanical operation of the component parts of organic matter, was not shaken.

The main object of Hooke, Grew, and Malpighi was to elucidate the fundamental phenomena of life, that is, the fabric and operation of organic matter. The goals of the two Dutchmen were less ambitious. Since Swammerdam was convinced of the ultimate inscrutability of nature he rejected the possibility of gaining any real insight into the mechanism of physiological processes. In his view the ultimate objective of scientific enquiry was the accurate description of life's manifestations. Consequently, he concentrated on the examination and careful description of the appearance and anatomy of insects. In so doing he disclosed a wealth of detail hitherto hidden from the naked eye or simply overlooked by investigators.

The bulk of Leeuwenhoek's investigations was also not primarily undertaken to explain how nature operates in animals and plants. Although Leeuwenhoek linked his earliest investigations with Cartesian-inspired explanations for the operation of organic matter, his grasp of Cartesianism was insufficient to carry through such a scheme for the whole realm of organic nature. Moreover, repeated investigations demonstrated that some of Leeuwenhoek's early explanations were mistaken because they were founded on inadequately resolved details. As a result, from a relatively early date in his career Leeuwenhoek ignored the explanation of his findings in the light of Cartesianism, and devoted his time entirely to the accurate description of organisms and organic structure.

Of these five microscopists, three—Hooke, Grew, and Malpighi—thus attempted to explain their observations from within the current philosophies of nature. Of these three Hooke's was undoubtedly the most ambitious undertaking, because he tried to substantiate with his observations the corpuscular worldview. Despite the fact that Hooke's presentation of his argument became very fragmentary as soon as it reached the texture and operation of the organic world, *Micrographia* appears to be a fairly well-documented exposition of the corpuscular conception of nature. The scope of Malpighi's work was much more limited than Hooke's, because Malpighi's aim was only to elucidate the origin of the phenomena of life. Despite the many gaps in his research Malpighi's published works cover both animal and vegetable physiology, to such an extent as to suggest a reasonably thorough discussion of both subjects, in particular because he stressed the basic similarity between the physiological processes in animal and vegetable bodies. Grew's work was even more limited in comparison with Hooke's because he dealt only with the vegetable world, but within his chosen field Grew presented a comprehensive and fairly convincing picture of the generation and operation of vegetable matter. Despite his catholic choice of subjects Leeuwenhoek's studies appear fragmentary in comparison with the investigations and writings of the other men, especially because he never attempted to gather his various investigations into one schema.

Reading through Hooke's *Micrographia* it is obvious that he regarded the mechanical and corpuscular philosophies as complementary to each other. Corpuscular philosophy pertained to the construction of matter, whereas mechanical philosophy was concerned with the operation of structured matter. Like Hooke, Grew discussed his views on vegetable structure and physiology in terms of the corpuscular and mechanical philosophies, and he applied these two terms in exactly the same way as Hooke. Although Malpighi was not concerned with the primary structure of matter, his writings on physiological phenomena are consistent with such a distinction between the corpuscular and mechanical philosophies. Even though there are only slight traces of corpuscular and mechanical philosophies in Swammerdam's and Leeuwenhoek's writings on microscopic subjects, the new way of thinking about nature certainly influenced both men. Swammerdam had been educated in Cartesian physiology during his student days in Leiden, but he came to regard its explanatory power as inadequate very soon after (or even before) he defended a Cartesian-inspired mechanism of respiration. Leeuwenhoek, on the other hand, attempted to explain (especially in the early years of his career) some of his findings in terms of a Cartesian physiology. However, as soon as he found that further investigation might amend his present knowledge to such an extent that his original explanations would become untenable, he abandoned this practice.

The approaches adopted by the five leading microscopists toward scientific investigation were very similar. Each stressed the crucial importance of careful observation, a sentiment best expressed by Hooke and Swammerdam, who both wrote that scientific enquiry must begin with the hands and the eye, and that any theories founded on subsequent observation must again be checked in nature. In fact only Swammerdam executed this scheme unconditionally because Hooke so favored the corpuscular hypothesis that he molded his observations to fit that thesis. The fact that the Royal Society played so prominent a role in the lives of four of the leading microscopists, that is, Hooke, Grew, Malpighi, and Leeuwenhoek, also illustrates the positive influence of Baconian motivation in science on the rise of microscopy.

Having analyzed and compared the substance of these men's microscopic investigations, we are left with the question of how far they influenced each other. Hooke, Malpighi, Grew, Swammerdam, and Leeuwenhoek certainly knew of each other's work, and from relevant remarks in their published works it is clear that they appreciated it. Although they investigated parts of the same subjects and reached comparable results, there appears to have been no exchange of viewpoint among them concerning the interpretation of their findings, either in person or in correspondence. The personal contacts between them were necessarily limited as they lived far apart, in England, the Low Countries, and Italy. Leeuwenhoek and Swammerdam met a few times at Leeuwenhoek's home, but since Swammerdam found Leeuwenhoek's learned intercourse barbaric, and Leeuwenhoek in his turn felt aggrieved about what he perceived as Swammerdam's underhanded use of his findings, these encounters were presumably not very productive.[1] Hooke and Grew, on the other hand, met quite regularly during the 1670s at the meetings of the Royal Society. Apart from the fact that Grew acknowledged, in the introduction to his *Comparative anatomy of trunks*, that Hooke encouraged him to employ a microscope for his exploration of plant structure, there are no indications of a serious exchange of ideas.

Most of the extant correspondence concerns business contacts. Grew and Hooke both wrote to Malpighi and Leeuwenhoek in their capacity as secretary of the Royal Society. Through these exchanges of letters news was communicated concerning the recent publications and investigations of third parties. Malpighi's knowledge of Leeuwenhoek's investigations, for instance, derived from such letters. The limitations of such a source of information, and the fragmentary dissemination of news generally at that time, are well illustrated by the fact that virtually all Malpighi learned about Leeuwenhoek's investigations was that the latter had discovered minute organisms in pepper water.[2] Since this subject was beyond the scope of Malpighi's interests it is understandable that he merely mentioned Leeuwenhoek's discovery to some of his correspondents but did not discuss it any further.[3] Still, it is surprising that Malpighi acknowledged no further awareness of Leeuwenhoek's investigations since he regularly received copies of the *Philosophical Transactions*. Even though he could not read English he must have recognized the substance of Leeuwenhoek's investigations from the illustrations.[4] However, even in cases where both writer and addressee had investigated the same subject, they did not discuss their respective results. Neither Hooke nor Grew, writing on behalf of the Royal Society to Malpighi and Leeuwenhoek, raised questions of their own in their correspondence; they only dealt with matters related to the printing of publications[5] or relayed such questions as were raised by the members in general.[6] In this way Grew and Leeuwenhoek corresponded about the latter's ideas concerning generation and particularly on his criticism of de Graaf's opinion on the function of the ovaries.[7]

One exception to the general rule that these men did not discuss their work is a short exchange between Grew and Leeuwenhoek concerning some details of the structure of plants. Typically, the opponents did not join in this discussion in person, but their points of view and questions were mediated through Henry Oldenburg, then acting secretary of the Royal Society. Leeuwenhoek sent a letter with descriptions of several specimens of wood to the Royal Society on 21 April 1676 and raised some specific questions to be put to Grew.[8] This letter was printed in the *Philosophical Transactions* of 18 July 1676 with anonymously published notes that came directly from Grew. Prior to publication a letter containing the same remarks was sent to Leeuwenhoek, who commented on them in his letter to the society of 29 May 1676.[9] Apart from some minor details Leeuwenhoek and Grew disagreed basically on two points.

The first point of debate was the valves within the wood vessels. Leeuwenhoek asserted that there were valves in the vessels, while Grew thought not. He considered that the observed protuberances within the wood vessels were artifacts produced by the knife slicing through the wood. The question of valves in the vessels was a much debated issue both within the circle of the Royal Society and the académiciens in Paris. At several meetings of the society Hooke devised and performed an experiment to settle the issue: he tried to force quicksilver through the length of a tree twig.[10] Similar experiments became rather popular throughout the scientific community in Europe. The van Musschenbroek workshop, for instance, sold a specially designed device as an accessory for an air pump for such experiments.[11]

The second point was that Grew described the medullary rays as composed from separate cells very much like those observed in the pith, whereas Leeuwenhoek regarded these as a third kind of vessel, running in a horizontal direction between the bark and the pith. Neither of the opponents changed his mind on these issues after hearing the other's point of view.

Reading this exchange between Leeuwenhoek and Grew it also becomes clear that communication between them was difficult because as yet there had been no development of a strict terminology with which to describe the details of the specimens they examined. Grew and Leeuwenhoek did not really understand one another because they were referring to different things. For instance, when Leeuwenhoek questioned Grew about the globules of which he thought the walls of the wood vessels were composed, he was alluding to what he regarded as the fundamental units of construction of organic matter, while Grew replied as though Leeuwenhoek had described what are now called the tyloses. Conversely, when Grew mentioned the fibers from which he thought the walls of the vessels were constructed he meant the elementary fibers of organic matter, while Leeuwenhoek thought he was referring to the thickenings on the walls of the spiral vessels.

Yet, despite the lack of personal contacts and the lack of information about each other's work, the results of one microscopist occasionally spurred another to renewed research, as in the following case. As soon as Swammerdam had read Malpighi's treatise on the silkworm, he set out to verify Malpighi's findings for himself. In so doing he "experienced something about the silkworms that Malpighi has not observed. I do not know whether I shall write to himself, or whether I shall have it printed; it concerns the heart, the medulla spinalis and the orifices of the bronchi."[12]

In the end Swammerdam, as mentioned in a previous chapter, did neither. He destroyed his notes on the silkworm in 1675, but asked Stensen to send the drawings he had made to Malpighi. Malpighi referred to these drawings in his autobiography and discussed the criticism by Swammerdam implicit in them.[13] One of Swammerdam's drawings depicted the brain and first thoracic ganglion of the silkworm larva. He wrote elsewhere that Malpighi had not seen the oesophagal nerve ring and had even overlooked the brain.[14] Malpighi acknowledged Swammerdam's critical remarks on his drawing of the nervous system, insofar as he admitted that he had overlooked one ganglion, but he indignantly denied that he had missed the brain.

A second drawing concerned the male reproductive organ. Swammerdam, as opposed to Malpighi, had discovered no passage between the seminal vesicles and the seminal ducts. Malpighi deemed this so serious that he reinvestigated the matter and found his earlier observations confirmed. In his autobiography he included a number of drawings illustrating the correctness of his previously published observations.[15] Malpighi and Swammerdam, therefore, stimulated each other to greater anatomical accuracy.

In sum, the five leading microscopists shared various incentives for applying the microscope to the organic world, especially the new philosophical framework and the Baconian style of scientific investigation. Yet, they received barely any intellectual inspiration from each other. This is perhaps not surprising as it appears that they found few faults with each other's observations and the interpretation thereof.

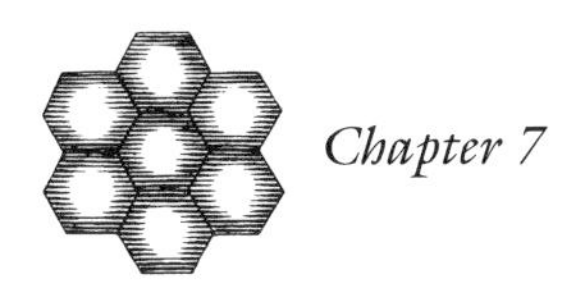

Chapter 7

Measuring the Impact of the Microscope

By counting and charting the number of books published on microscopy—as well as the number of articles on the subject that appeared in three different journals—it has become clear that Hooke had assessed correctly, in both 1664 and 1691, the contemporary situation in microscopy. His initial statement was made at a time when the number of publications began to rise suddenly. His confidence therefore seems to have been shared by a number of serious investigators, who published the results of their microscopic studies. By the time Hooke made his second statement, the number of publications on microscopy had sunk below the peak reached in the first half of the 1680s. In the succeeding years the decline persisted, and by the time of Leeuwenhoek's death in 1723 almost no one else published anything on the subject of microscopy. Thus the few decades in the second half of the seventeenth century that saw Hooke's, Malpighi's, Leeuwenhoek's, Swammerdam's, and Grew's celebrated discoveries were indeed a unique period in the history of microscopy.

A review of the main events in the history of the microscope demonstrates that the development of this instrument began to gain momentum only after the microscopic exploration of nature had led to some widely appreciated discoveries. It is therefore apparent that the rise of microscopy stimulated the development of the microscope rather than the other way around (even though the latter is a feasible and attractive explanation for the microscope's increased use). The demands of contemporary research influenced the design of microscopes, highlighted the advantages of the single-lens microscope, and prompted the improvement of the mechanical construction of the microscope. Although many makers of microscopes proposed and effected various modifications of the optical system, the performance of microscopes was not significantly improved during the second half of the seventeenth century. However, the development of the microscope had until well into the eighteenth century hardly any bearing on the reach of microscopic discovery. This point is well illustrated by comparing the merits of descriptions of the structure of the compound eye of arthropods given by several investigators between 1625 and 1704. As far as detailed knowledge of the structure of the compound eye is concerned, Odierna and Swammerdam could not be bettered. Odierna's research of 1644 far surpassed many of the later investigations by Hooke, Catelan, Puget, and others. In fact, the research of some of these later investigators yielded descriptions that were definitely inferior to Odierna's findings, and were only slightly better than Stelluti's description of more than half a century earlier. Therefore, the microscopes of 1644 were certainly capable of resolving as much detail as many instruments produced at a much later date.

The present enquiry confirms that the earliest microscopic investigations were indeed presented by their respective authors within the context of a mechanistic interpretation of the phenomena of nature. Such a concern is apparent both in Odierna's discussion of the eye of the fly and in Borel's comments on some of his one hundred microscopic observations. These men referred to both mechanical actions and the corpuscular basis of matter while commenting on their findings.

The men in the group known as the "Oxford physiologists," as well as the members of the newly founded Royal Society, explicitly appreciated the microscope within the context of the corpuscular hypothesis of matter. Several of them, among them Walter Charleton, had already examined a number of familiar objects with the microscope some time before Hooke magnificently surpassed their investigations in his *Micrographia*. Even though their excursions into the microworld were brief and rather casual they served to develop their perception of the intricacy and efficiency of the organic world. The microscopic findings were not in themselves important for these men, because the factual demonstration of minute organic structures only provided them with a starting point for their reasoning in support of the corpuscular nature of matter. Consequently their efforts did not develop into full-scale investigations. The quintessence of their argument was that, if the processes of life could be so efficiently performed by such minute organs as those observed in small insects, the corpuscular nature of matter could hardly be denied.

Hooke's *Micrographia* was also inspired by the corpuscular philosophy. The most striking characteristic of this magnificent book is the enthusiasm with which it demonstrates the advantages to be gained by employing a microscope when studying small objects. The fact that in the early 1660s two men, Henry Power and Robert Hooke, independently published their explorations of the microscopic world with the blessing of the Royal Society, of which both were members, might suggest that the contemporary ambience of the society spurred these publications. In both books the society's philosophy, the Baconian program of inductive reasoning in science, is very prominent. In Hooke's case the members of the Royal Society greatly influenced the progress of his microscopic pursuits, and eventually they licensed the publication of *Micrographia*. Even though Hooke, who had been appointed their Curator of Experiments, could probably not entirely ignore the members' directive that they should be presented with more and more views of the microscopic world, he met their demand with such enthusiasm that an inspiration transcending the pressures of the Royal Society obviously encouraged his exploration of this subject. Henry Power's *Experimental Philosophy* bears the same stamp of enthusiastic investigation and appreciation of scientific instruments as Hooke's *Micrographia* and has often been regarded as a perfect reflection of the buoyant mood during the early years of the Royal Society. However, it has been established that Power executed his various researches far away from, and largely independent of, the Royal Society.[1]

Therefore, it was not so much the Royal Society as a body that influenced the publication of two microscopic works within so short a timespan, but rather the rising momentum of Baconianism in England, reinforced (as far as Hooke and Power were concerned) by the impact of the corpuscular theory of matter. From their books it is apparent that the union of the empirical method of scientific investigation with a rational explanation of the phenomena of nature, whether on a corpuscular basis or a mechanical one, transformed microscopic investigation from a disjointed series of casual observations into a sustained effort.

The microscope came into use simultaneously within both the circles of the Royal Society and a wholly different scientific setting, among anatomists. Many anatomists of the second half of the seventeenth century concentrated their investigative efforts primarily on the fabric of the organs of the human body, with a view to understanding its operation. Some of these investigators employed the microscope to a greater extent—Malpighi, for instance—or to a lesser extent to explore the minute machinery of the organs in a variety of living beings. The instrument was thus used as a means to aid novel physiological investigations. The aspirations of the anatomists active in this field of enquiry were stimulated, if not prompted, by Descartes's reductive explanation of familiar physiological phenomena, which amounted to an acute appreciation of the physical properties of organic structures as the basis of physiological processes. Whereas Descartes reasoned largely on the basis of well-known data, the anatomists who were inspired by his approach applied their knives and other anatomical means to investigate the fabric of organic matter more precisely before attempting a mechanical explanation of physiological processes. Therefore, these men also combined a fresh conception of organic matter with the experimental approach.

Thus to summarize, the microscopic observations published during the 1660s and early 1670s are deeply concerned with the issues raised by the corpuscular philosophy and Cartesian-inspired physiology. They also reveal a deep commitment to the experimental method on the part of the investigators. The beginning of the heyday of microscopy in the seventeenth century was therefore the result of a combination of the novel seventeenth-century conception of nature with the experimental method. Even though both these innovative developments in science originated several decades earlier, it is not surprising that the upsurge in microscopic investigation occurred no earlier than approximately 1660. Both the new explanatory framework and the experimental method had to mature and gain acceptance beyond the circle of their initial proponents before they could be applied in other areas than those in which they originated (i.e., the physical sciences).

Despite the fact that Hooke's *Micrographia* was enthusiastically received,[2] no scientist except Henry Power modeled his research on Hooke's example. This is perhaps not surprising since his efforts, despite their superficial success, ultimately failed to demonstrate the point he had wanted to prove. Indeed, neither Hooke's microscopic investigations, nor Power's for that matter, supported or contributed anything substantial to the corpuscular philosophy. Hooke speculated freely on his ideas concerning the ultimate structure of matter, the processes that lead to the organization of matter into material objects, and some of the phenomena of living things such as the growth of very simple plants on the basis of the corpuscular philosophy, but his observations did not support those speculations. Neither did his findings throw any further light on the processes of life, on which the interest of contemporaneous microscopists was focused. Hooke was only superficially interested in physiological phenomena. When he declared, in the lecture of 1693 mentioned earlier, that with the microscope the "internall mechanism" of animate bodies might be discovered,[3] he had in mind such processes as the propulsion of recently ingested blood through the intestines of a louse rather than the machinery within a gland producing a specific juice from the blood. By that time physiological theorizing had transcended such a rather primitive level of enquiry, being concerned with the elemental structure of organic matter. Nevertheless, his magnificent and accurate illustrations of vegetable and animal specimens were highly praised and occasionally copied by authors active in the field of natural history, such as Buonnani.

The anatomists who were concerned with the animal "oeconomy," on the other hand, had revealed many structural details and proposed theories as to the function of these particulars. Malpighi had discovered round bodies in the brain, kidney, and spleen and interpreted these as glands, the "string of pearls" structure of the muscles had been observed by several investigators and fitted into the current "explosion" theories of muscular contraction, and the alveolar capillary blood vessels had been related to several feasible functions of the lungs. Hooke's hope that the microscope would reveal "the subtility of the composition of the Bodies, the structure of their parts, the various texture of their matter, the instruments and manner of their inward motions"[4] seemed thus to have been amply fulfilled.

Even though the resolution of gross organic texture into minute structures formed the common denominator in the researches of the "mechanic physiologists" they varied widely in their explanations of the final causes of the various phenomena of the animal "oeconomy," as is evident from a comparison between the writings of Willis and Malpighi. Willis, like Hooke and Power, was a corpuscularist who attributed the dynamics of the physiological processes to the interactions between various kinds of particles. Consequently Willis explained the microscopic details of organic structures within his scheme of the fermentations and ebullitions that resulted from these interactions. In his view, therefore, organic structures were dependent on the chemical processes. Malpighi, on the other hand, thought that the very physical properties of the texture of organs brought about the various physiological phenomena, and illustrated his ideas with purely mechanical, graphic similes, such as levers, filters, and sieves. In his view the arrangement of the constituent parts of the various organs was the cause of the phenomena.

Leeuwenhoek joined in the detailed examination of the fabric of organic matter from a very early date in his career. He did not work within a well-defined program like Malpighi and Willis, and his microscopic researches were not embedded, as theirs were, in traditional anatomical methods. In fact, the microscope was virtually the only tool he used in his research. In the early years of his career Leeuwenhoek usually interpreted his observations from within the custom and idiom of Cartesian physiology. Most of the physiological mechanisms he proposed on the basis of his observations were derived from his original globule theory of organic matter and operated purely mechanically.

The initial microscopic data were certainly incorporated, or at the very least were taken into consideration, in contemporary physiological theories. The microscopic data were exploited to perfect existing gross mechanical processes, for instance, the filtration of urine and the contraction of muscles. Malpighi's globular bodies in the kidney elaborated Bellini's earlier theory concerning the mechanism of urine production, and Leeuwenhoek's original description of the structure of muscle fibers fitted neatly into a refined mechanical scheme of muscle operation as a result of an "explosion" within the muscle structure.

However, in some cases microscopic data pointed to facts that were incompatible with current physiological theory, as was the case with the nerves. Microscopic research revealed no canals or pores within the nerves although other experimental evidence seemed to point to the transmission of some kind of substance through the sensory and motor nerves, a nervous fluid, animal spirit, or similar entity. Nevertheless the hollow-nerve theory was maintained throughout the eighteenth century. The very outcome of microscopic research played an important part here, because it had amply demonstrated that visible objects could be resolved into smaller structures, and in some cases these smaller entities appeared to be composed of still smaller parts. Therefore, the fact that a certain structure eluded the efforts of the microscopists did not have to mean that it did not exist; it might merely mean that it was too small to be detected with the instruments at one's disposal. Balancing the available microscopic data concerning the construction of the nerve against prevailing theory, the former simply did not carry enough weight to overthrow the latter; neither did the data suggest an alternative mechanism.

The development of the rationalizations offered by Leeuwenhoek at the end of his observations reflects the plight of microscopists in this respect. Initially, when still cherishing his globule theory, Leeuwenhoek offered various physiological mechanisms on the basis of these globules.

However, as soon as more sophisticated observations (both in terms of technique and experience) demonstrated that instead of globules, fibers occurred throughout organic nature as the fundamental unit of construction, only a few of his Descartes-inspired explanations were upheld, while most disappeared from his writing. Indeed, causal explanations disappeared almost completely from them.

In the wake of the microscopic exploration of organic matter for the sake of the corpuscular philosophy or the animal "oeconomy," the microscope began to be used increasingly in natural history as well, thereby continuing the feeble start Cesi, Stelluti, and Peiresc had made. Many men, among them Leeuwenhoek, Swammerdam, Tyson, Buonnani, and Huygens, investigated the habitus and anatomy of a fairly large number of different species of minute organisms.

The microscopic work of some of these microscopists was very firmly placed within a well-defined program of research, but the work of others completely lacked such a theoretical framework. Swammerdam's microscopic research, for instance, originated from his ambition to demonstrate—in opposition to the prevailing opinion—that insects are highly organized beings with a distinct manner of development. Having demonstrated his point in his *Bloedeloose dierkens* of 1669, he became passionately absorbed in the anatomy of insects and collected a wealth of materials transcending the immediate needs of his original program. Tyson performed his microscopic investigations in the service of a program devoted to compiling a comprehensive natural history of animals. There is a sharp contrast to these men in such figures as Griendel von Ach and Buonnani, both of whom collected a series of superficial observations. They were mainly concerned with recording the striking particulars of minute organisms, a preference that met with wide approval to judge by the praise voiced by various reviewers, such as Sachs and Schrader.

Among the wide range of small and smaller animals from which they could choose, the microscopists showed a definite preference for insects and infusoria. Consequently, through the investigations of men like Swammerdam, Leeuwenhoek, and Muralt, a more detailed description of the exterior of many kinds of insects, some of which had already been studied by Borel and Hooke, was accumulated. The dissection and description of the interior of the insect body was almost exclusively the province of three of the leading microscopists: Malpighi, Leeuwenhoek, and—outshining even these colleagues—Swammerdam. The anatomy of insects presented a delightful surprise in that it underscored the extreme delicacy of organic structure, but it did not affect the basic ideas concerning animal form or function. The anatomy of insects conformed essentially to the familiar plan observed in the perfect animals.

The investigation of the microorganisms, even though the discovery of these minute, active organisms delighted the eyes of the observers, posed several problems. As far as could be seen they possessed little internal structure, although, judging from the terminology employed in their description, such structures were definitely thought to be present. Indeed, reasoning from the analogy that was simultaneously being reinforced as a result of the microscopy of insects, the performance of the principal functions of the infusoria was attributed to minute structures comparable to the ones observed in larger animals. Yet it was even more frustrating that the mode of their reproduction remained obscure, despite novel techniques of enquiry, so that very unsatisfactory theories had to be put forward.

Despite the numerous widely appreciated discoveries that the microscope contributed to the corpus of knowledge of living bodies, Hooke gave an accurate assessment of the current situation when he declared in 1691, in a lecture to the Royal Society, that besides Leeuwenhoek few men now undertook microscopic investigations. Judging by the careers of the five leading men, the decline of microscopy had already set in by 1680. Hooke did not publish any more microscopic investigations after *Micrographia;* his *Microscopium* is essentially an essay in microscopic techniques. Malpighi published his last work containing solid microscopic research in 1678. Grew published his final book, in which his microscopic researches (published and unpublished alike) were accumulated, only four years later. Swammerdam had died early in 1680, and the manuscripts he left behind were not to be published until half a century after his death. Only Leeuwenhoek was still active. Indeed, by comparison it seems as if he had just begun: for the next thirty years he was to send a continuous stream of letters to the Royal Society and other correspondents, in which he described his various observations. However, most of his epoch-making discoveries were made in the first half of his career as a microscopist: that is, before 1690. His researches after that date consisted of a great deal of confirmation, repetition, and elaboration of his earlier work. Since the men of lesser importance, such as Griendel and Tyson, did not publish any substantial number of books or papers between them, the output of microscopic research dropped.

All the same, the resolution of the lenses of late-seventeenth- and early-eighteenth-century instruments was certainly adequate for a further exploration of the fine fabric of nature. Pierre Lyonet's anatomy of the caterpillar of the goat moth,[5] Otto F. Müller's inventory of infusoria,[6] and Johannes Hedwig's description and elucidation of the reproductive organs of mosses and ferns[7] bear ample testimony to that assertion.

The apparent decline of microscopic research resulted from the nature of the issues tackled and the ensuing results. In the course of the 1670s and 1680s it became clear that the common result of the microscopists' joint efforts was that an overwhelming number of fibers and vessels constituted the principal structural element throughout organic nature. Be it kidney, muscle, or plant, if these were studied through the microscope, vessels invariably met the eye. A muscle appeared to be constructed from smaller muscle fibers and a nerve from smaller nervous fibers, ad infinitum. It was thus established that organic matter was composed of a number of rather uniform units: the vessels and fibers. Even though they appeared to be so simple and so similar, these units made up the distinctive fabric of totally different organs and appeared to be able to bring about a wide range of different phenomena. Therefore, the central assumption of the anatomical method, that anatomical data point to physiological function, was found to be inappropriate in microscopic investigation. Consequently a different kind of analysis was needed. In view of the universal presence of fibrous elements Descartes's original procedure for creating a mechanical physiology, viz., simply reinterpreting the familiar physiological explanations of the sixteenth and early seventeenth century in terms of simple physical causes corresponding with the characteristics of the various organs, seemed inadequate.

Toward the end of the seventeenth century the main question in physiology, how the several processes are brought about by the fabric of vessels constituting the organs, was approached from various angles. The Newtonian physiologists in England, headed by Archibald Pitcairne, initiated a mathematical analysis of the properties of the diverse vascular systems in the organs. They attributed the various physiological phenomena to the different courses, diameters, and ramifications of the vessels. Herman Boerhaave, on the other hand, effected a deductive analysis of organic structure starting from the elemental fiber, and linked the various stages of organization of the fibers to organic structure and physiological function.[8] In both cases an intensified examination of the texture of organs was immaterial to the progress of the analysis. Ironically, the starting point for these analyses was to a large extent derived from microscopic observations, whereas the development of subsequent theories had no need of advanced microscopic research.

Thus, in physiology the theories of the first generation of mechanical physiologists appeared to be based on mechanical similes that bore no relationship to the uncovered fabric of the organs. However attractive such mechanical similes seemed to be for the explanation of physiological processes, because of the expressive, easily understandable picture they presented to the mind, they could not be reconciled with the evidence of the microscope. The inevitable conclusion was that the modus operandi of physiological phenomena must be of greater finesse, a conclusion strengthened by the experience that gross organic structures could be thought to operate as the result of several distinct mechanical causes, only to produce identical results. A case in point is the difference between the explanations offered by Malpighi, Grew, and Perrault for the transportation of fluids through the large vessels in plants. Malpighi summoned the temperature of the air as the final cause, while Grew favored capillary force combined with the pressure of cells next to the vessels, and Perrault assumed that the continuous bending of branches in the wind constituted the moving force for the transportation of liquids in plants.

However right Hooke may have been in thinking that so much more might be done with the microscope, the conceptual foundation of physiology toward the end of the seventeenth century had no immediate need of the instrument. The investigations in the closing decades of the century had given rise to a definite theory of the construction of organic matter, in which the fiber formed the essential unit, both in the vegetable and animal world. Although simple mechanisms like the lever and sieve were incompatible with the fabric of fibers and vessels, such a fabric agreed with the mechanical conception of nature. Indeed, this fabric, an intricate arrangement of vessels of various diameters, was thought to bring about the various physiological processes as a result of the fluids that circulated through the vessels. Such a system appeared an attractive subject for mathematical analysis, but did not need further ocular investigation. In a sense then, the goals of the early microscopists formed the basis for the decline in microscopy some decades later.

In natural history the situation was very different. During the second half of the seventeenth century the appearance of many different minute organisms had been accurately described for the first time, thanks to the microscope, and a host of new organisms had been discovered. Yet it must be acknowledged that the natural historians often did not aspire to the same high level of microscopic achievement as their predecessors of the seventeenth century had attained. René-Antoine Réaumur, for instance, used an optical instrument with rather weak lenses to study the habitus, habitat, and habits of various species of small insects thoroughly. However, when he found it necessary to describe their anatomy he simply relied on the published findings of the early microscopists and did not attempt to dissect any specimens personally. Thus he copied Malpighi's drawing of the ovaries of the silk moth and Swammerdam's dissection of the ovaries of the bee.[9] Since the object of Réaumur's attention was the external particulars of different species, which could be adequately studied with the help of weak lenses, he did not have much use for powerful microscopes. In that respect Réaumur was typical of the majority of his contemporary colleagues in natural history. The use of the microscope was profitable but certainly not essential as far as the scientific concerns and achievements of contemporary natural history—the principles of classification and distinction of species-characteristics—were concerned. Consequently, even though the microscope contributed to the development of natural history and systematics, its contribution is rarely mentioned or appreciated in historical surveys of microscopy. By the same token, most of the books and articles in which the results of eighteenth-century microscopic research were published would not, on the basis of the criteria applied in the first chapter of this book, have been considered as publications relevant to microscopy.

In his lecture of 1691 Hooke had complained that the microscope now served only for "diversion and pastime." Indeed, the eighteenth century was to become the heyday of the popularization of science, but its roots lay in the preceding century. The popularization of science went hand in hand with natural theology, a way of thinking that appreciated the delicately wrought, but highly efficient, constructions in nature as strong arguments for premeditated design, proving the existence of God. The quintessence of such arguments is already apparent in the writings of the seventeenth-century microscopists. In his treatises Grew frequently pointed to the intricacy of the fabric of vegetable matter as a reminder of the magnificence of God, but the overwhelming presence of God's hand in the delicate structure of organic matter was most forcefully articulated by Swammerdam in his *Biblia naturae*.

Looking back over the achievements of the five leading microscopists and their peers in the second half of the seventeenth century one cannot but agree with Hooke: "Among the various methods of inquiring into the latent and internall structure and composition of Bodys, that by the help of Microscopes has not been the least significant . . . when we consider the nature of the informations it has and may yet afford, yt may possibly appear to Deserve to be ranked among the most considerable."[10]

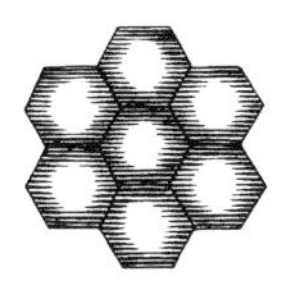

Conclusion

THE UPSURGE in the number of books and articles published on microscopy in the second half of the seventeenth century represents a burst of activity in that field of research that lasted for some four decades. Analysis of the contents of these publications indicates that microscopic investigation in the 1660s was largely inspired by the rising influence of the mechanical philosophy in the life sciences. Initially the delicate construction of minute organisms as revealed by the microscope was employed simply as an argument in support of the mechanical basis of matter. At that time, in the 1650s and 1660s, scholars, including both physicians and physicists, envisioned the operation of physiological processes in terms of simple, everyday mechanical actions. Early microscopic investigations, such as Robert Hooke's examination of the sting of the stinging nettle, revealed sundry structures that could easily be explained with reference to such mechanical actions and therefore appeared to support the mechanical explanation of the operation of organic matter.

The development of the compound microscope, both as regards the mechanical construction and optical system, was greatly advanced by the widespread appreciation of the scientific community of microscopic discoveries such as Malpighi's description of the capillaries in the lungs. Prior to 1660, when the microscope was only occasionally employed for the investigation of nature, the construction of the microscope remained essentially unchanged for some forty years. From 1660 onwards numerous proposals for the improvement of the optical system and the introduction of a considerable number of different types of microscopes coincide with the flourishing of microscopy. Although the magnification and resolution of the microscopes current toward the end of the seventeenth century was not vastly superior to earlier specimens, the later instruments were more versatile and much easier to handle.

With these instruments an abundance of microscopic particulars throughout animate nature were revealed, ranging from the structure of muscles, brains, and the stems of plants to the minute organisms living in staggering numbers in the smallest pools of water. As the details of the fabric of an increasing number of different organs and tissues were investigated it became evident that organic matter was extremely fine and complexly structured. Moreover, it appeared that comparable structural components recurred throughout the animal and vegetable world, a fact that prompted a novel notion of the organization of organic matter, that is, vegetable and animal matter was understood to be constructed from millions of extremely delicate vessels and fibers.

The operation of the finely woven fabric of organic matter did not correspond with the familiar mechanical actions that were initially put forward to explain its operation. Consequently, a more sophisticated explanatory scheme was developed, a theory that proposed that the actions of the various organs were brought about by the different courses, diameters, and ramifications of the capillary vessels within the organs. This theory relied heavily on the data that had accumulated in the preceding years and appeared not to need further confirmation with the microscope. Therefore, the microscope could be dispensed with as far as the "animal oeconomy" was concerned.

However, toward the end of the seventeenth century the microscope had also become established as a fairly regular tool in anatomy and natural history. It was actively employed in both fields of research throughout the eighteenth century. In a sense the decline of microscopy is therefore a delusion created by the circumstance that the microscope in the second half of the seventeenth century was primarily employed in a subject, that is, the workings of the body, on which the attention of many scholars was focused and that changed considerably in outlook because of that same tool, whereas in natural history the microscope was used to describe the appearance of small animals and played only a minor and inconspicuous role in that field.

The body of microscopic investigations in the seventeenth century was carried out by only five men: Robert Hooke, Jan Swammerdam, Marcello Malpighi, Nehemiah Grew, and Antoni van Leeuwenhoek. The last named stands out in particular when the sheer bulk and scope of his inquiries are considered. A number of these celebrated men's peers contributed their share, a share that by comparison appears rather unassuming. Hooke, Malpighi, Swammerdam, Grew, and Leeuwenhoek worked fairly isolated from each other, but it is remarkable that all had ties of varying degrees with the Royal Society in London. On occasion the Royal Society actively promoted microscopic investigation, while the Académie des Sciences in Paris was, as a body, much less interested in the microscope.

The five dominant microscopists did not share similar aspirations. On the one hand, Hooke, Grew, and Malpighi were primarily concerned with the elucidation of the fundamental phenomena of life from a mechanistic point of view. Leeuwenhoek and Swammerdam, on the other hand, merely aspired to accurately discover and describe the structure of organisms, which in Swammerdam's case meant almost exclusively insects, or organic substances in general. However, they did share an intense regard for careful observation and great diligence in finding ways and means to apply a new and as yet almost unexplored instrument on a virgin field of research.

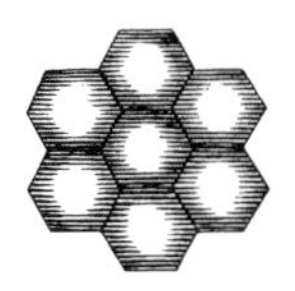

Appendixes

A. Books on Microscopy Published between 1625 and 1750

Year	Author	Title
1625	Cesi and Stelluti	Apiarium
1630	Stelluti	Persio tradotto*
1644	Odierna	L'occhio della mosca
1646	Fontana	Novae coelestium *
	Kircher	Ars magna lucis*
1656	Borel	Observationum microscopicarum centuria
1661	Malpighi	De pulmonibus
1664	Power	Experimental philosophy*
1665	Hooke	Micrographia
	Malpighi	De externo tactus organo
	Malpighi	Epistolae anatomicae cum de omento
1666	Malpighi	De viscerum structura
1669	Malpighi	De bombyce
	Swammerdam	Historiae generalis insectorum
1673	Grew	The anatomy of roots
	Malpighi	De formatione pulli in ovo
1674	Willis	Pharmaceutice rationalis I*
1675	Malpighi	Anatomes plantarum I
	Malpighi	De ovo incubato
	Willis	Pharmaceutice rationalis II*
	Swammerdam	Ephemeri vita*

Year	Author	Title
1675	Grew	The anatomy of trunks
1676	Sturm	Collegium experimentale*
1678	Hooke	Cutlerian lectures*
1680	Swammerdam	Biblia naturae* (manuscript)
1681	Schrader	Dissertatio epistolica
1682	Grew	The anatomy of plants*
1683	De Heide	Anatome mytuli
1686	Leeuwenhoek	Ontledingen en ontdekkingen*
1687	Bonomo and Cestoni	Osservazioni intorno
	Griendel, von Ach	Micrographia nova
	Leeuwenhoek	Vervolg der brieven*
1688	Leeuwenhoek	Den waarachtigen ommeloop des bloeds*
1689	Leeuwenhoek	Natuurs verborgentheden ontdekt*
1691	Buonanni	Micrographia curiosa
1693	Leeuwenhoek	Derde vervolg der brieven*
1694	Leeuwenhoek	Vierde vervolg der brieven*
1696	Leeuwenhoek	Vijfde vervolg der brieven
1697	Leeuwenhoek	Zesde vervolg der brieven
	Malpighi	Opera posthuma*
1698	Bidloo	Brief aan Antony van Leeuwenhoek
1706	Puget	Observations sur la structure des yeux
1718	Joblot	Descriptions et usages*
	Leeuwenhoek	Sendbrieven*
1734	Cuno	Durch dessen verfertigte microscopia
1743	Baker	An attempt at a natural history
1744	Trembley	Mémoirs pour servir d'un genre de polype
1745	Lieberkühn	De fabrica villorum intestinorum
	Needham	An account of some new microscopical discoveries
	Parsons	The microscopical theatre of seeds

* Partly included in Chart 1.

B. Articles on Microscopy in the *Journal des sçavans* between 1665 and 1750

Issue	Pages	Author	Title
1678	187–88	De la Hire	Nouvelle découverte des yeux de la mouche
1680	167–70	Catelan	Observations touchant les deux parties des insectes
1681	92–94	Catelan	Autres observations
	140–42	Catelan	Suite des observations
1704	102–23	Puget	Observations sur la structure des yeux de divers insectes

Note: All issues of the journal between 1665 and 1750 were examined; only those that contained reports of microscopic investigations are listed here.

C. Articles on Microscopy in the *Philosophical Transactions* between 1665 and 1750

Issue	Pages	Author	Title
2 (1667)	425–28	E. King	Concerning Emmets, or ants
4 (1669)	1043–47	E. King	Concerning the organs of generation
8 (1673)	6037–38	Leeuwenhoek	A specimen of some observations
	6116–18	Leeuwenhoek	The figures of some microscopical observations
9 (1674)	23–25	Leeuwenhoek	More microscopical observations
	121–28	Leeuwenhoek	Concerning blood, milk, . . .
	128–31	Leeuwenhoek	About sweat, fatt, teares
	178–82	Leeuwenhoek	More observations
10 (1675)	378–80	Leeuwenhoek	Concerning the optic nerve
	380–85	Leeuwenhoek	About the texture of the blood

Issue	Pages	Author	Title
11 (1676)	653–60	Leeuwenhoek	Concerning the texture of trees
12 (1677)	821–33	Leeuwenhoek	Concerning little animals
	899–905	Leeuwenhoek	Of the carneous fibres
	1002–5	Leeuwenhoek	Of teeth and hair
	1040–46	Leeuwenhoek	De natis è semine genitali
13 (1683)	74–81	Leeuwenhoek	An abstract of a letter
	113–44	E. Tyson	Lumbricus latus
	154–61	E. Tyson	Lumbricus teres
	197–208	Leeuwenhoek	Concerning the appearance of several woods
	347–55	Leeuwenhoek	About generation by an animalcule of the male seed
14 (1684)	566–67	Grew	The description and the use of the pores in the skin
	568–74	Leeuwenhoek	About animals in the scurf of the teeth
	586–92	Leeuwenhoek	Concerning scales within the mouth
	780–89	Leeuwenhoek	About the cristallin humour of the eye
15 (1685)	883–95	Leeuwenhoek	Concerning the parts of the brain
	963–79	Leeuwenhoek	Concerning the salts of wine and vinegar
	1073–90	Leeuwenhoek	Concerning the various figures of salts
	1120–34	Leeuwenhoek	Concerning generation by an insect
	1156–58	G. Garden	Concerning the proboscis of bees
	1236–38	W. Molyneux	Concerning the circulation of the blood
16 (1686–91)	506–10	E. Tyson	Lumbricus hydropicus
17 (1693)	593–94	Leeuwenhoek	The abstract of two letters
	646–49	Leeuwenhoek	Concerning animalcules found on the teeth
	700–708	Leeuwenhoek	Concerning the seeds of plants

Issue	Pages	Author	Title
	754–60	Leeuwenhoek	Several observations on cinnabar and gunpowder
	838–43	Leeuwenhoek	On the texture of the bones
	861–65	E. King	On the animalcula, in pepper-water
	949–60	Leeuwenhoek	Observations on the seed of cotton
18 (1694)	194–99	Leeuwenhoek	The history of the generation of an insect
	224–25	Leeuwenhoek	Concerning the differences of timber
19 (1695–97)	254–59	J. Harris	Of vast numbers of animalcula
	269–80	Leeuwenhoek	On eels, mites
	280–87	S. Gray	Several microscopical observations
	790–99	Leeuwenhoek	Concerning the eggs of snails
20 (1698)	169–75	Leeuwenhoek	Concerning the eyes of beetles
	376–78	B. Allen	An account of the scarabaeus Galateus pulsator
21 (1699)	301–08	Leeuwenhoek	Concerning the animalcula in semine humano
22 (1700–1701)	447–55	Leeuwenhoek	Concerning the circulation and stagnation of the blood
	509–18	Leeuwenhoek	Concerning the worms in sheeps livers
	552–60	Leeuwenhoek	Concerning the circulation and globules of the blood
	635–42	Leeuwenhoek	Concerning worms
	659–72	Leeuwenhoek	Concerning some insects
	673–76	F. Poupart	Concerning the insect called Libella
	739–46	Leeuwenhoek	On the animalcula in semine masculino
	786–92	Leeuwenhoek	Concerning excrecencies growing on willow leaves
	821–24	Leeuwenhoek	Concerning the spawn of codfish

Issue	Pages	Author	Title
	867–81	Leeuwenhoek	Concerning spiders
	899–903	Leeuwenhoek	Concerning the causes of different tastes of water
	903–7	Leeuwenhoek	Concerning several microscopical observations
23 (1702–3)	1137–51	Leeuwenhoek	Concerning the animalcula in semine masculino
	1152–55	Leeuwenhoek	On rain water
	1177–1201	W. Cowper	Divers schemes of arteries and veins
	1296–99	G. Bonomo	Concerning the worms of humane bodies
	1304–11	Leeuwenhoek	Concerning green weeds growing in water
	1357–72	C.H.	Some microspocal observations
	1386–93	W. Cowper	Concerning the cure of aposthumation of the lungs
	1430–43	Leeuwenhoek	On some animalcula in water
	1461–74	Leeuwenhoek	Concerning the seeds of oranges
	1494–1501	C.H.	Two letters
24 (1704–5)	1522–27	Leeuwenhoek	Concerning the worms observed in sheeps livers
	1537–55	Leeuwenhoek	Concerning the figures of sand
	1614–28	Leeuwenhoek	Concerning Cochineel
	1723–30	Leeuwenhoek	Concerning the flesh of whales
	1730–40	Leeuwenhoek	Concerning the tubes . . .
	1740–48	Leeuwenhoek	Concerning tobacco ashes
	1774–84	Leeuwenhoek	Concerning some fossils
24 (1705)	1784–93	Leeuwenhoek	Concerning animalcula on the roots of duck-weed
	1794–97	Leeuwenhoek	On staining the fingers with a solution of silver
	1843–55	Leeuwenhoek	Concerning the barks of trees
	1868–74	Leeuwenhoek	On the seed-vessels and seeds of Polypodium

Issue	Pages	Author	Title
	1906–17	Leeuwenhoek	Concerning the figures of the salts of crystal
	2158–63	Leeuwenhoek	On the pumice-stone
25 (1706–7)	2205–9	Leeuwenhoek	On the seeds of several East-India plants
	2305–12	Leeuwenhoek	On the structure of the spleen
	2416–24	Leeuwenhoek	Of the salts of pearls
	2425–32	Leeuwenhoek	Concerning the particles of silver dissolved in aqua fortis
	2446–55	Leeuwenhoek	On the cortex Peruvianus
	2456–62	Leeuwenhoek	Concerning the whiteness on the tongue in fevers
26 (1708–9)	53–58	Leeuwenhoek	On the blood vessels and membranes of the intestines
	111–23	Leeuwenhoek	Upon the tongue
	126–34	Leeuwenhoek	Upon red coral
	210–14	Leeuwenhoek	Upon the white matter on the tongues of feverish persons
	250–57	Leeuwenhoek	Concerning the circulation of the blood in fishes
	294–301	Leeuwenhoek	On the palates of oxen
	416–19	Leeuwenhoek	Upon the hair . . .
	444–49	Leeuwenhoek	On the particles of crystalized sugar
	479–84	Leeuwenhoek	Upon the configuration of diamonds
	493–98	Leeuwenhoek	Upon the edge of razors
	499–502	Leeuwenhoek	Upon the same subject as the former
27 (1710–12)	20–23	Leeuwenhoek	Upon the chrystalized particles of silver
	24–27	A. Adams	Concerning the manner of making microscopes
	316–20	Leeuwenhoek	Upon the animalcula in semine of young rams

Issue	Pages	Author	Title
	398–415	Leeuwenhoek	Upon the production of mites
	438–46	Leeuwenhoek	Upon the seminal vessels, muscular fibres and blood of whales
	518–22	Leeuwenhoek	Upon the contexture of the skin of elephants
	529–34	Leeuwenhoek	Upon muscles
28 (1713)	160–64	Leeuwenhoek	On the animalcula found upon duckweed
29 (1714–16)	55–58	Leeuwenhoek	Concerning the fibres of the muscles
	59–61	W. W. Muys	Concerning the frame and texture of the muscles
	486–90	R. Bradley	Observations relating to the motion of sap in vegetables
	490–92	R. Bradley	Some microscopical observations . . . on the vegetation . . . of moldiness
31 (1720–21)	91–97	Leeuwenhoek	Upon the bones and the periosteum
	129–34	Leeuwenhoek	Upon the membranes enclosing the fasciculi of fibres
	134–41	Leeuwenhoek	Upon the vessels in several sorts of wood
	190–99	Leeuwenhoek	On the muscular fibres of fish
	200–203	Leeuwenhoek	Upon the seeds of plants
	231–34	Leeuwenhoek	De osculis
32 (1722–23)	72–75	Leeuwenhoek	Concerning the muscular fibres
	93–99	Leeuwenhoek	Concerning the particles of fat
	151–56	Leeuwenhoek	Upon a foetus . . . of a sheep
	156–61	Leeuwenhoek	Upon the callus of the hands and feet
	199–206	Leeuwenhoek	De particulis et structura adamantum
	400–407	Leeuwenhoek	De structura diaphragmatis
	436–37	Leeuwenhoek	De globulis in sanguine
	438–40	Leeuwenhoek	De generatione animalium

Issue	Pages	Author	Title
41 (1739–41)	276–88	J. Baster	On the worms which destroy the piles
41 (1740)	448–55	H. Baker	The discovery of a perfect plant in semine
	725–29	H. Miles	Concerning the circulation of the blood
	770–75	H. Miles	Concerning the seed of fern
42 (1742–43)	283–91	A. Trembley	Observations and experiments upon the freshwater polypus
	416–19	H. Miles	Observations on the mouth of the eels in vinegar
	422–36	H. Baker	Some account of the insect called the freshwater polypus
	616–19	H. Baker	On a polype dried
	634–41	J. T. Needham	Concerning . . . malm
43 (1744–45)	35–36	H. Baker	Concerning a new discovered sea-insect
	169–83	A. Trembley	Upon several newly discovered species of freshwater polypi
44 (1746–47)	60–66	J. Hill	Concerning the manner of the seeding of mosses
	67–69	J. Sherwood	Concerning the minute eels in paste being viviparous
	150–58	R. Badcock	Observations on the farina foecundans of the Holyoak
	189–91	R. Badcock	Concerning the farina foecundans of the yew-tree
	627–55	A. Trembley	Upon several species of small water insects of the polypus kind
45 (1748)	615–66	J. T. Needham	Observations on the generation . . . of animal and vegetable substances

Note: All issues of the journal between 1665 and 1750 were examined; only those that contained reports of microscopic investigations have been listed.

D. Articles on Microscopy in the *Miscellanea curiosa* between 1670 and 1750

Issue	Pages	Author	Title
Decade 1			
1 (1670)	34–49	Ph. J. Sachs	Messis observationum microscopicarum è variis authoribus collectarum
2 (1671)	53–55	J. Paterson Hain	Experimenta microscopica
	192–93	J. Paterson Hain	Cochlea in lucii capite reperta
4–5 (1673–74)	68–69	J. Paterson Hain	De vermibus è stomacho rejectis
8 (1677)	81–87	L. Schröck	De animali moschifero
Decade 2			
1 (1682)	71–74	Chr. Mentzel	De Muscis quibusdam Culiciformibus, pediculosis, grylliformis et aliis
	136–37	J. von Muralt*	Anatomia Pediculi
	137–38	J. von Muralt*	Anatomia Pulicis vulgaris
	138–39	J. von Muralt*	Pulex florum scabiosae
	139–41	J. von Muralt*	Anatomia Crabronis
	141–42	J. von Muralt*	Anatomia Cimicis murorum & lignorum *Holtzwentelen*
	142–47	J. von Muralt*	Examen anatomicum Grylli sylvestris
	148–50	J. von Muralt*	Scarabeiaei majalis foliacei anatome *Laubkäffer*
	154–56	J. von Muralt*	Gryllo-talpa
	156–58	J. von Muralt*	Scarabaeus liliaceus *Herrgottskühlein*
	158–62	J. von Muralt*	Anatomia Muscae vulgaris
	166–67	J. von Muralt*	Scorpius
	167	J. von Muralt*	Cicindela

Issue	Pages	Author	Title
2 (1683)	40–42	J. von Muralt*	De Locusta viridi majore
	42	J. von Muralt*	De Locusta viridi majore alia
	43–44	J. von Muralt*	De Blattâ, *Brodkäffer*
	44	J. von Muralt*	De Forficula, *Ohrenmügeler*
	44–46	J. von Muralt*	De Cantharide I. *Goldkäffer*
	46–47	J. von Muralt*	De Cantharide viridi chlorochryso
	47–48	J. von Muralt*	Cantharis miniata
	48–49	J. von Muralt*	Papilio flavus
	49	J. von Muralt*	De Violâ lutea cum Phryganio
	50	J. von Muralt*	De Pulice campestri, *Herdflöhe*
	50–51	J. von Muralt*	De Perla Ribesiorum
	51–52	J. von Muralt*	Aranea vulgaris
	52–55	J. von Muralt*	Bombyx
	58–59	J. von Muralt*	De Gryllo-talpa
	162–63	M. Tiling	De Vermium sub herbis putrefactis generatione
	189–94	J. von Muralt*	De insectis
	194–95	J. von Muralt*	Phryganion Perlae
	195–97	J. von Muralt*	Anatomia Perlae, *Augenschießer*
	197–98	J. von Muralt*	Phygolampis lacustris. *Glyßling*
	198–200	J. von Muralt*	Squilla mollis lacustris
	200–201	J. von Muralt*	Examen Papilionis vulgaris albi
	201–3	J. von Muralt*	Consideratio Scarabaei Cornuti, Maris et foemellae
	203–5	J. von Muralt*	Erucarum anatome
	205–6	J. von Muralt*	Scarabaeus Rosarum argenteus, *Rosenkäfferle*
	206–7	J. von Muralt*	Scarabaeus vaginipennis subsultans: *Knecht stand uff*
	264–67	D. Spielenberger	De Vermibus nivalibus
	295–97	Chr. Mentzel	Musca pulex vel cimex
	297–98	Chr. Mentzel	Papilio-Blatta alis plumosis

Issue	Pages	Author	Title
3 (1684)	117–23	Chr. Mentzel	De perlis praestantissimo Muscarum genere
4 (1685)	96–98	G. S. Polis	De vermibus vomitu rejectis
	98–100	G. S. Polis	De Muscis polonicis exoticis
5 (1686)	323–24	G. W. Wedel	Sanguis per microscopium examinatus
6 (1687)	285	G. Hann	Sabulum urinarium, in microscopio examinatum
	333–36	J. M. Hoffmann	De foetu monstroso
	459–65	J. G. Volckamer	Anatomia cervae
7 (1688)	294	G. Hann	Rorismarini fabrica, in microscopio examinata
	468–69	J. G. Volckamer	De folio salviae glandifero
	480–81	L. Schröck	De verme quadrato
10 (1691)	411–15	L. Schröck	De fungulis minimis seminiferis
Appendix	33–44	J. C. Bonomo	Observationes circa humani corporis teredinem
Decade 3			
1 (1694)	52–53	J. Lanzonus	De lactis examine ope microscopii facto
	126–27	G. C. Gahrliep	De vermiculo erucae simile per urethram excreto
2 (1694)	161–67	S. Reisel	De baccis seu granis floribus Quercus Adnatis
3 (1695/96)	3–4	Ph. Fraundorffer	De partu Millepedarum
	291–93	D. Nebel	De Glandula lachrymali Harderiana
5–6 (1697–98)	70–72	R. J. Camerarius	De Vermibus nivalibus
	220–21	D. Nebel	De Vermiculis plumbum
7–8 (1699–1700)	51	J. C. Bautzman	De Verme in lapide reperto
	256–57	G. C. Gahrliep	De minutiis vegetabilibus curiosis
	325–26	U. Staudigel	De musca, compluribis vermiculis foeta

Issue	Pages	Author	Title
Continued as *Ephemerides*			
1–2 (1712)			
Appendix	153–65	A. Vallisnieri	De ovario anguillarum
3–4 (1715)			
Appendix	137–46	S. Löber	Epistola [De Locustis]
7–8 (1719)	338–47	J. J. Dillenius	Hirudinibus et duobus Papilionibus
9–10 (1722)			
Appendix	484–508	F. Scufoni	Observationes circa Locustas
Continued as *Acta physico-medica*			
2 (1730)	270	Ph. H. Pistorius	Pilis cum ovulis insecti insoliti in Ceraso
3 (1733)			
Appendix	49–59	D. Nissolius	De origine et natura Kermes
4 (1737)	14–20	D. Revegli	De culicum generatione
8 (1748)	51–53	J. Baster	De generatione pilorum in corpore Humano

Note: All issues of the journal between 1665 and 1750 were examined; only those that contained reports of microscopic investigations have been listed.

* In chart 1 I have reduced the count of Muralt's contributions by half because of the extreme shortness of some of the contributions and because it is often unclear whether he used a microscope for these investigations.

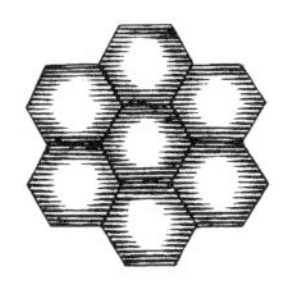

Notes

Introduction

1. Hooke (1665), Preface, fourth page.
2. Hooke (1726), p. 261.
3. For a review see Turner (1972).
4. This point is discussed by Rooseboom (1953); A. R. Hall (1963), pp. 166–67; van Helden (1983); Meinel (1988); and C. Wilson (1988).
5. E.g., Henry Baker's *The microscope made easy* and George Adams's *Micrographia illustrata, or, the knowledge of the microscope explain'd.*
6. Nicolson (1956).

Chapter 1: A New Instrument Appraised

1. C. Singer (1914, 1915); Belloni (1969). The original source is J. Wodderborn, who wrote in his *Quator problematum* of 1610 that he had heard Galileo relate of a microscopic observation to a friend.
2. Van Helden (1977b); Turner (1985).
3. The letter is preserved in Middelburg, Rijksarchief in Zeeland, MSS "Staten van Zeeland," 1633, f. 31r. The letter is transcribed and translated in van Helden (1977b), pp. 35–36.
4. Govi (1889).
5. Disney (1928), p. 99.
6. A transcription of Peiresc's description of his microscope (dated ca. 1622) was published by Humbert (1951). Isaac Beeckman's drawing of a microscope by Drebbel was published by de Waard in his edition of Beeckman's *Journal* (1939–53), 3:442.
7. Van Helden (1983).
8. Rooseboom (1967); van Helden (1977a).
9. Hooke (1665), Preface, twenty-second page.
10. Divini (1668).

11. Birch (1756–57), 2:348–49.

12. Griendel (1687), pp. 3–4.

13. Huygens (1888–1950), *Oeuvres complètes* (hereafter *OC*), 5:308–9, 318, 330–31; letters dated 5, 10, and 17 April 1665, respectively.

14. Hooke (1678), p. 313.

15. Huygens, *OC*, 13:675.

16. Divini (1668).

17. Griendel (1687), pp. 5–6.

18. Van Cittert (1934a); van der Star (1953); Bradbury (1967b); Bracegirdle (1978b).

19. Van Zuylen (1981a, 1981b).

20. Data calculated by van Zuylen (1981a, 1981b) from Henry Baker's account of Leeuwenhoek's microscopes in the *Philosophical Transactions* of 1740.

21. Van Seters (1933).

22. Hevelius (1673–79), 1:303–4, plate S.

23. Huygens, *OC*, 8:205.

24. Bedini (1963).

25. On 16 and 30 July and on 20 August; see Huygens, *OC*, 8:96 n. 1 and 8:112 n. 2, respectively.

26. For a detailed analysis see Fournier (1989).

27. Butterfield issued a pamphlet in 1679 entitled *L'usage du nouveau microscope, fait avec une seule et très petite boulle de verre* depicting this model. A variant of this design is preserved in the Science Museum in London; a Chapotôt microscope is preserved in the *Istituto e Museo di Storia della Scienza di Firenze* in Florence; a Depouilly microscope is preserved in the Science Museum in London. At some time before 1700 Cuno published a pamphlet concerning his microscope of this type entitled *Bericht an die Herren Liebhabere Optischer Kunst-Wercken den Handgriff und Gebrauch deren hierbey verziechneten Microscopiorum.*

28. Quoted and translated in Clay and Court (1932), p. 10, from J. Wodderborn's *Quator problematum* . . . (1610).

29. Belloni (1967), pp. 129–30.

30. Humbert (1951) contains a transcription from which this quotation was taken: "les plus petits paroissoient aussy bien formés que les plus gros" (p. 156). For Peiresc see also H. Brown (1974).

31. Quoted in Govi (1889), n. 8, from a letter by Peiresc to Girolamo Aleandreo dated 7 June 1622.

32. Quoted in Kidwell (1970), pp. 107–8, from Stelluti (1630).

33. Beeckman (1939–53), 2:240, manuscript note dating from 1623.

34. See his introduction to Thomas Moufet's *Insectorum sive minimorum animalium Theatrum* of 1634.

35. In his *Pseudoxia epidemica; or, Enquiries into very many received tenets and commonly presumed truths,* p. 176.

36. Fontana (1646), pp. 148–51; Kircher (1646), pp. 834–35.

37. For information on Odierna, see Abetti (1974); D. W. Singer (1956); Pighetti (1961); Fodera Serio et al. (1983).

38. Odierna (1644), p. 19: "Vadi il perito Anatomista con l'estremità d'un Coltellino di somma sottigliezza (à questo officio apparecchiato) mentre l'Occhio suo sú l'Orificio superiore dello strumento molto avvicinato, sta con accorezza mirando, vadi incidendo la superficie Cornea, sin che dalla continuta sustanza, la separi."

39. E.g., Charleton (1654), p. 166.

40. In his *Plus ultra: or, The progress and advancement of knowledge since the days of Aristotle,* p. 66.

41. In his *Iatrologismorum seu medicinalium observationum pentecostae quinque utilibus praeceptis,* pp. 131–33.

42. Hughes (1958).

43. For Borel see Farber (1970); Carré (1974); Chabbert (1968).

44. Borel (1656), in the dedication: "Quo atomi visibiles quasi, et minuta insectula in molem Colosseam transmutantur; cujus ope, in atomis illis animatis partes innumerae deprehenduntur, novaeque phisicae fores indies aperiuntur, adeò, ut Dei Majestas magis in hisce parvis Corporibus, quam in giganteis, elucescat, eorumque; stupenda compages Atheississimos etiam convincat, eosque ad supremi eorum Architecti notitiam, admirationem & venerationem deducat."

45. Borel (1656), p. 16: "pedes, nervos, oculos & omnes partes animalis."

46. Ibid., pp. 38–39.

47. Ibid., p. 36: "Cor, renes, testiculi, jecur, pulmo, aliaque corporis parenchymata, plexum esse organularum & fibrarum videbis, ceu criba, quibus variae substantiae à natura secernuntur, juxta pororumfiguras, quibus certis certae figurae tantus atomis datur ingressus."

48. Webster (1967).

49. Sachs (1670); Buonanni (1691).

50. Hooke (1665), pp. 112–13.

51. Grew (1682), p. 9.

52. Malpighi (1665), *De lingua;* Belloni (1965c).

53. Cole (1921).

54. Malpighi (1666), *De renibus;* Belloni (1965d).

55. J. R. Baker (1945); van der Pas (1970).

56. Leeuwenhoek (1718), p. 101, letter dated 21 August 1714; van der Pas (1970).

57. Muys (1714); van der Pas (1970).

58. Malpighi (1666), *De cerebro cortice;* Belloni (1966b, 1968a).

59. Clarke and Bearn (1968).

60. Lindeboom (1975c), p. 98.

61. Leeuwenhoek (1939 ff.), *Collected Letters* (hereafter *CL*), 8:7–57, letter dated 7 September 1688; Palm (1978).

62. Leeuwenhoek, *CL,* 1:191–203, letter dated 4 December 1674.

63. Hooke (1678), p. 310.

64. Specimens mounted on microscopes were sent by Leeuwenhoek to the Royal Society shortly before his death in 1723 and were described by Folkes (1723) and later by H. Baker (1740). In 1745 Leeuwenhoek's microscopes were auctioned, many still fitted with specimens; see Bracegirdle (1984).

65. Adelmann (1966), pp. 833 and 2336–37; Malpighi (1675–79), *De ovo incubato.*

66. Fournier (1989), p. 582.

67. See, e.g., Henry Baker (1742), chap. 14, and George Adams (1746), chap. 10.

68. Hooke (1665), Preface, twenty-second page.

69. Van Seters (1933); Schierbeek (1950–51), 1:90–119.

70. Balthasar (1710), chap. 10.

71. Jurin (1718), pp. 761–62.

72. Kisch (1951).

73. Hooke (1665), Preface, sixteenth page.

74. Zanobio (1960, 1971); Belloni (1962, 1985).

75. Kerckring (1670), p. 177: "Scisne centrum istius visionis esse minutissimum?, scis colores ita variare, ita rebus hac videndi ratione affundi, ut quis earum nativus & verus sit, dijudicari nequeat? scis denique ea, quae sic interpolatis visionibus percipiuntur, fieri posse, ut discreta appareant, verè unita & continua quae sunt?"

76. Chérubin d'Orleans (1677), Preface, sixth page: "tres alterées, disproportionnées, & difformes, avec des additions & des diminutions purement conjecturales."

77. Malpighi (1697), pp. 31–32, quoted in Adelmann (1966), 1:370.

78. Hooke (1665), Preface, twenty-fourth page.

79. Frederik Ruysch, for example, categorically rejected Malpighi's contention that the cortex of the brain consisted of glands: see especially his polemic with Herman Boerhaave originally published in 1722 as *Opusculum anatomica de fabrica glandularum in corpore humano,* of which a Dutch translation entitled *Ontleedkundige Verhandeling over het Maakzel der Klieren* is included in Ruysch (1744), part 3, pp. 1151–1228.

80. Paraphrase of Chérubin d'Orleans's argument on the seventh and eighth pages of the Preface to *La vision parfaite* (1677).

81. Fournier (1985) contains a list of Bidloo's microscopic illustrations and depicts some of them.

82. Dennis (1989).

83. Malpighi (1669), *De bombyce;* quoted in Cole (1944), p. 183.

84. A. R. Hall (1966); Harwood (1989).

85. Ornstein (1938); Kronick (1962); Houghton (1975).

86. Quoted in Houghton (1975), p. 14.

87. Kronick (1962), p. 77.

88. These sources are Nelson (1902); Catalogue (1929); Nachet (1929); de Martin (1973).
89. See, e.g., the nineteenth chapter of *Cerebri anatome.*
90. Lindeboom (1975c).
91. Comparable results are reported in Nowak (1984), p. 5.
92. C. Singer (1914).

Chapter 2: The Leading Microscopists

1. For full particulars see Keynes (1960).
2. Hooke (1705); Hooke (1726).
3. Espinasse (1956); Westfall (1972); Shapin (1989).
4. Shapin (1989).
5. Boas Hall (1991).
6. Hooke (1665), Preface, seventh page.
7. Ibid., Preface, fourth page.
8. Ibid., p. 171.
9. Hooke (1726), p. 172.
10. Hooke (1665), Preface, fourth page.
11. Hooke, quoted in Birch (1756–57), 3:364–65.
12. Bennett (1989).
13. Hesse (1966); Oldroyd (1972, 1987).
14. Hooke (1665), Preface, fifteenth page.
15. Ibid., p. 1.
16. Ibid., pp. 125–31.
17. Ibid., p. 130.
18. Ibid., p. 127.
19. Ibid.
20. Ibid., p. 154.
21. Ibid., pp. 11–22 and 82–87, respectively; for a discussion of Hooke's "property of congruity" see Henry (1989).
22. Hooke (1665), p. 151.
23. Ibid., p. 152.
24. Hooke (1705), pp. 71–148.
25. Ibid., p. 142.
26. Hooke (1665), pp. 123–25, 132–34.
27. Ibid., p. 194.
28. For information on Malpighi, see Adelmann (1966); Belloni (1974).
29. Respectively, Glisson's *Anatomia hepatis* (1654); Willis's *Cerebri anatome* (1664); Stensen's *De musculis et glandulis observationum specimen* (1664); and Duverney's *Traité de l'organe de l'ouie* (1683).
30. Malpighi (1697), pp. 1–102.

31. T. S. Hall (1969), 1:342–48; Settle (1970); Balaguer Perigüell (1971).

32. Cavazza (1980).

33. Adelmann (1966), 1:533–88.

34. See ibid., 1:558–64, for a paraphrase of Sbaraglia's arguments.

35. Dewhurst (1958); Wolfe (1961).

36. Quoted in Young (1929), p. 7, from *De pulmonibus* (1661).

37. This expression was introduced by Belloni in his various writings on Malpighi.

38. Quoted in Belloni (1975), from Malpighi's autobiography in *Opera posthuma*.

39. Quoted in Adelmann (1966), 1:574, from *Riposta* in Malpighi's *Opera posthuma*.

40. Quoted in Young (1929), p. 19, from *De pulmonibus* (1661).

41. Quoted in Adelmann (1966), 2:955, from *De formatione pulli in ovo* (1673).

42. Quoted in T. S. Hall (1951), pp. 152–55, from *De renibus* (1666).

43. See Adelmann's (1966) translation (1:309) of paragraphs from *De liene* (1666) and his paraphrase (1:300) of *De cerebri cortice* (1666), respectively.

44. See his introduction to *De viscerum structura* (1666); for a paraphrase see Adelmann (1966), 1:296–98.

45. Quoted in Adelmann (1966), 1:570, from *Riposta* in *Opera posthuma*.

46. Borelli (1680–81), iii. See also Adelmann (1966), 1:150.

47. Adelmann (1966), 1:355, quotes Malpighi's letter to J. T. Schenk, dated 20 November 1670.

48. Quoted in Hayman (1925), pp. 259–60, from *De renibus* (1666).

49. Boerhaave (1737); Schierbeek (1947); Lindeboom (1975c, 1982a); Winsor (1976).

50. Borch (1660–65), pp. 269–72, 299; Nordström (1955).

51. Lindeboom (1975c), p. 54.

52. Swammerdam was offered this sum by the Duke of Tuscany, but declined because the offer entailed that he come and take care of the collection in Italy; Lindeboom (1975c), p. 72. See Swammerdam (1681) and Lindeboom (1980) for the later history of the collection.

53. Swammerdam (1737–38), p. 394: "Siet, soo oververwonderlyk is GOD, ontrent deese kleene Beeskens, soo dat ik durf seggen, dat ontrent de Insecten GODS onnmoemelyke wonderen versegelt syn, ende dewelke segelen zig komen te openen, als men het boek der Natuur, de Bybel van Natuurelyke Godsgeleertheid, en waar in GODS Onzienelykheid sigtbaar wort, neerstig komt te doorbladeren; want schatkameren van onnoemelyke wonderen openbaaren haar alsdan; en de verborgene Schepper wort in deese kleene Dierkens soo openbaar, dat de ondervindingen omtrent deselve, my voor de allergrootste bewysen dienen, om syne eeuwige Goddelykeheid ende Voorsienigheid, tegens alle syne ontkenners onversettelyk te bewysen."

54. Swammerdam (1675), Preface, p. 5: "Ick heb nu langh genoegh mijn tijt ende arbeyt besteedt, in het ondersoecken van de natuur, ende mijn verdurve eyge wil ende behaagen daar in gevolght. Waarom ick nu voorneem de wille Gods alleen te volgen; mijn wil aan hem over te geeven; ende alle mijne gedachten van de meenighvuldigheeden af te trecken, om die alleenigh aan hemelsche bedenckingen op te offeren."

55. Swammerdam (1675), p. 87.

56. Belloni (1968b).

57. Lindeboom (1975c, letters 11 up to and including 40; 1982a).

58. Swammerdam (1737–38), p. 386: "alleen tot GODS lof, ende sonder eenige andere de minste insigt te vervolgen."

59. As argued by Bäumer (1987).

60. Schulte (1968); Scherz (1976).

61. Swammerdam (1737–38), p. 837; Schulte (1968).

62. Cole (1938b); Lindeboom (1975a).

63. Ruysch, *Dilucidatio valvularum in vasis lymphaticis, et lacteis.*

64. Blasius (Blaes), *Anatome medullae spinalis et nervorum inde provenientium.*

65. This engraving was dedicated to Nicolaas Tulp and appeared, slightly changed, as one of the illustrations in *Miraculum naturae.*

66. Birch (1756–57), 3:41, 94; Lindeboom (1973), pp. 118–19.

67. Swammerdam (1737–38), p. 868.

68. Visser (1981) discusses this point in some depth.

69. Swammerdam (1737–38), pp. 868–69: "Indien onse reeden valsch ende gebrekkelyk is, indiense niet door de ondervindingen kan ondersteunt werden, daar door beweesen werden, ende in de selve eyndigen, soo dunkt ons, dat'er geen sterker nogte krachtiger reedenen kunnen weesen; als dewelke uyt de ondervindingen ende de ervarentheeden selfs, daarse in moeten eyndigen, gehaalt werden. Synde alle andere reedenen, dewelke deese vaste ende onbeweeglyke grontvest niet en hebben, op hoe veel optellingen ende besluyten sy ook steunen, enigsints voor verdagt te houden; ende soo se met de ondervindingen niet overeen en koomen, geheel te verwerpen." This passage was originally published in *Bloedeloose dierkens* (1669).

70. Swammerdam (1737–38), pp. 870–71: "Gelyk als we van alle saaken geen waaragtige ondervindingen kunnen verkrygen, ende alsoo geen klaar ende onderscheydentlyk begrip van deselve hebben (als van die, dewelke voor ons gesigt te kleen, ende van andere, dewelke daar te ver afgeleegen syn) soo moeten wy ook ons niet dwaselyk inbeelden, van oyt door onse reeden tot de waare ende eygentlyke kennis van de oorsaaken dier dingen, ik laat staan tot die van haare waaragtige uytwerkingen, te sullen koomen. Leggende . . . onse aldergrootste Wysheid, niet in de kennis van de oorsaken der dingen, maar alleen in een net ende onderscheydentlyk begrip van derselver waare vertooningen, ofte haare

uytwerkingen . . . geleegen." This passage was originally published in *Bloedeloose dierkens* (1669).

71. Swammerdam (1737–38), p. 503: "Want oog, hant, verstant en instrumenten, syn daar al te saam om haar groote kleenheid te onvermogent toe."

72. Swammerdam (1667), chap. 1, paragraph 11.

73. Swammerdam (1737–38), pp. 835–60; Nordström (1955); Lindeboom (1975b).

74. Swammerdam (1737–38), p. 664: "O GODT, uwe Werken syn ondoorsoekelijk, en alles dat wy daar van weten, of weten kunnen, syn niet als de doode schaduwen van de schaduwen der schaduwen uwer aanbiddelijke en ondoorsoekelijke werken; waar voor alle de verstanden der Menschen, hoe spitsvondig sy syn, moeten stomp worden, en haar domme onwetendheid bekennen."

75. Swammerdam (1737–38), p. 15: "de regelen ende orderen, van den alwysen Maaker, geheel onveranderlijk in den aard der saaken gestelt." This passage was originally published in *Bloedeloose dierkens* (1669).

76. Swammerdam (1737–38), p. 4.

77. Ibid., p. 37.

78. As Schierbeek (1947) and Winsor (1976) did.

79. Visser (1981); Ruestow (1985).

80. Lindeboom (1975c). In his letters to Thévenot Swammerdam, he referred to his "great work," in which these various investigations were to be brought together.

81. Bowler (1971); Ruestow (1985).

82. For the 1668 date see Lindeboom (1975c), Introduction, p. 13. For the possibility of the 1662 date see Borch (1660–65), p. 241; Nordström (1955).

83. Quoted in Winsor (1976), from Swammerdam (1737–38).

84. Swammerdam (1737–38), p. 34. This passage was originally published in *Bloedeloose dierkens* (1669).

85. Cole (1930), pp. 41–44; Needham (1934), pp. 148–49; Roger (1971), pp. 334–35.

86. Swammerdam (1737–38), p. 728.

87. Bowler (1971) discusses the distinction between the two.

88. Ruestow (1985).

89. Quoted in Bowler (1971), n. 41, from Swammerdam (1669).

90. Swammerdam (1737–38), p. 666: "En als voor een aflegging van de oude deelen, een nieuwe schepping, of een opstanding van het oude lichaam in een nieuw te agten is."

91. Ibid., p. 815.

92. As advocated by Bowler (1971) and Ruestow (1985), respectively.

93. Swammerdam (1737–38), p. 713: "Men met waarheid kan seggen, dat GODT maar een eenig Dier geformeert heeft, en dat onder oneyndige gestalten, buygingen, samen windingen en uytrekkingen van leedemaaten verborgen, en on-

derscheyden heeft: waar by hy dit selve een verschilligen aart, manier van leeven, en voetsel heeft toe geschikt."

94. Carruthers (1902); Arber (1940–41, 1942); Metcalfe (1972); Zirkle (1965); Bolam (1973).

95. Grew (1682), Preface, first page. The dates of the various lectures he presented to the Royal Society are mentioned in the *Anatomy of plants* (1682) and also in Birch (1756–57), vol. 3.

96. The year 1672 appears on the title page, but the book was ready just before the end of 1671, as it was presented to some members of the Royal Society on 7 December 1671; see Birch (1756–57), 2:984.

97. Grew (1672), Preface.

98. Ibid.

99. Birch (1756–57), 2:490, 3:47.

100. Hunter (1982); see also Birch (1756–57), 3:42, 47.

101. Birch (1756–57), 3:49.

102. Ibid., 3:359.

103. His thesis was entitled *De liquore nervorum*. Some years later he lectured to a meeting of the Royal Society on the same subject; see ibid., 3:228.

104. Metcalfe (1972); Hunter (1982).

105. Hunter (1982).

106. Grew (1682), Preface, fourth and fifth pages.

107. Birch (1756–57), 3:55, 56.

108. Ibid., 3:72, 176, 195, 228, 359.

109. Ibid., vol. 3, e.g., pp. 426–27.

110. Hunter (1982), n. 66.

111. Birch (1756–57), 4:344, lists the names of the new council, installed 1 December 1684.

112. Ibid., 4:121.

113. Ibid., 2:295, 301, 311, 468, 476, 477.

114. Ibid., 2:480.

115. Grew (1682), p. 143.

116. T. S. Hall (1969), 1:279–94; Hoppe (1976), pp. 226–27.

117. Grew (1682), Preface, first page.

118. Ibid., Idea, p. 3.

119. Ibid., p. 80.

120. Grew (1701), p. 1.

121. Grew (1682), p. 221.

122. Ibid., p. 239.

123. Ibid., p. 228.

124. Ibid., Idea, p. 2.

125. Ibid., Idea, p. 9.

126. Ibid., Idea, pp. 24, 5.

127. Hunter (1982), p. 195, transcription of a letter from Grew to Henry Oldenburg.

128. Birch (1756–57), 3:333.

129. *The Comparative Anatomy of Stomachs and Guts* (1681), which appeared by way of an appendix to *Musaeum regalis societatis.*

130. For a full discussion see Delaporte (1982).

131. Grew (1682), Dedication, second page.

132. Ritterbush (1964), pp. 74–76; Hoppe (1976), pp. 266–68; Delaporte (1982), pp. 37–42.

133. Grew (1682), p. 117.

134. Ibid., Idea, p. 4.

135. Bolam (1973), p. 228.

136. Grew (1701), p. 18.

137. Read on 10 December 1674 to the Royal Society and initially published with another lecture in *Experiments of Luctation* and later included in the *Anatomy of Plants.*

138. Grew (1682), p. 238.

139. Ibid., lectures read to the Royal Society in March 1676 (p. 255) and 21 December 1676 (p. 261), respectively.

140. Grew (1682), p. 160.

141. Ibid., p. 86.

142. Ibid., p. 87.

143. Ibid., p. 157.

144. Heniger (1973).

145. Leeuwenhoek, *CL,* 1:43, letter dated 15 August 1673.

146. Palm (1989), p. 160.

147. Dobell (1932); Schierbeek (1950–51); Heniger (1973); Castellani (1973).

148. Schierbeek (1950–51), pp. 46–59; Leeuwenhoek (1693), Dedication.

149. Halbertsma (1843); van Charante (1844); Haaxman (1875); Harting (1876); Beyerinck (1913); Dobell (1923, 1932); Schierbeek (1950–51); Castellani (1973); Lindeboom (1982b); Ruestow (1983).

150. Baas Becking (1924).

151. Meyer (1938); Cole (1937); Dobell (1932); Schierbeek (1950–51).

152. Van Berkel (1982).

153. Particularly by Hartsoeker in his "Extrait critique des lettres de feu M. Leeuwenhoek" which was added to his *Cours de physique.*

154. Dobell (1923), p. 43, quotes a letter in Constantijn Huygens to Robert Hooke, dated 8 August 1673.

155. Lindeboom (1975c), p. 108.

156. Huygens, *OC,* 10:52, letter by Leibniz dated 20 (30) February 1691.

157. Dobell (1932), pp. 40–41, quotes the letter by de Graaf to the Royal Society, dated 28 April 1673.

158. E.g., the letter of 9 October 1676 on the infusoria; see Birch (1756–57), 3:338, 346, 349, 352.

159. This is suggested by Heniger (1973).

160. Van Berkel (1982); see also Leeuwenhoek, *CL,* 2:171, letter dated 7 November 1676.

161. Birch (1756–57), 4:6.

162. Daniël van Gaesbeek; see Schierbeek (1950–51), pp. 44–46.

163. Van Berkel (1982).

164. Quoted in van Berkel (1982), from a letter by Leeuwenhoek dated 25 July 1707.

165. Leeuwenhoek, *CL,* 12:305, letter dated 9 June 1699.

166. Van Berkel (1982).

167. Dobell (1932); Schierbeek (1950–51); Rooseboom (1959); Heniger (1973).

168. Egerton (1968).

169. Leeuwenhoek, *CL,* 6:35, letter dated 2 April 1686.

170. Ibid., 5:229, letter dated 13 July 1685.

171. Ibid., 4:192–94, letter dated 28 December 1683; see also Snelders (1982).

172. Dobell (1932), pp. 73–76; Schierbeek (1950–51), p. 67; Heniger (1973).

173. Leeuwenhoek, *CL,* 1:331 and 3:385, letters dated 20 December 1675 and 3 March 1682, respectively.

174. Ibid., 2:367–69 (concerning bone and dentine) and 3:385–87 (concerning the texture of muscle), letters dated 31 May 1678 and 3 March 1682, respectively.

175. Ibid., 10:35–47; Garden also published a review of contemporary research and theories concerning generation in *Philosophical Transactions* 16 (1690–91), 474–83.

176. Ibid., 10:35–61, letter dated 19 March 1694.

177. Ibid., 2:33, letter dated 29 May 1676.

178. Ibid., 1:49–51, letter dated 15 August 1673.

179. Ibid., 1:31 (moulds), 75 (hair), 77 (nails), 113 (plants): letters dated 28 April 1673, 7 April 1674 (hair and nails), and 1 June 1674, respectively.

180. Ibid., 1:199, letter dated 4 December 1674.

181. Ibid., 1:87 (color of bone and teeth), 105 (opacity of tissues), 191 (color of the iris): letters dated 24 April 1674, 1 June 1674, and 4 December 1674, respectively.

182. Ibid., 1:56–60, letter dated 15 August 1673.

183. Ibid., 1:307, letter dated 14 August 1675; see also Snelders (1982).

184. Schierbeek (1939), (1950–51), pp. 149–50; Snelders (1982).

185. Leeuwenhoek, *CL,* 1:211, 279, letters dated 22 January 1675 and 26 March 1675; Huygens, *OC,* 7:400, letter dated 30 January 1675.

186. Leeuwenhoek, *CL,* 1:181, letter dated 9 October 1674.

187. Ibid., 2:369, letter dated 31 May 1678.

188. Ibid., 2:309, 3:255, 3:287; Leeuwenhoek (1702), p. 233: letters dated 14 January 1678, 14 June 1680, 12 November 1680, and 9 July 1700, respectively.

189. Leeuwenhoek, *CL,* 3:397, letter dated 3 March 1682. On crystals see *CL,* 1:311, 2:125, 5:21: letters dated 14 August 1675, 14 November 1679, and 5 January 1685, respectively.

190. Ibid., 3:55, letter dated 20 May 1679, and 5:25, letter dated 5 January 1685.

191. Ibid., 3:21, letter dated 25 April 1679.

192. Leeuwenhoek (1721), p. 136; letter dated 9 January 1720.

193. Leeuwenhoek, *CL,* 2:391, letter dated 27 September 1678.

194. Ibid., 3:57–63, letter dated 20 May 1679.

195. Ruestow (1984).

196. Leeuwenhoek, *CL,* 7:5–37, 8:277–91, 9:207–59, 11:179–217: letters dated 6 August 1687 (calander), 7 March 1692 (weevil and grain moth), 15 October 1693 (flea), and 20 February 1696 (louse), respectively.

197. Ibid., 3:329, letter dated 12 November 1680.

198. Ibid., 10:269–301, 9:67–77, letters dated 10 July and 20 August 1695.

199. Leeuwenhoek (1718), p. 286, letter dated 5 November 1716; (1702), p. 106, letter dated 23 June 1699; ibid., p. 282, letter dated 26 October 1700.

200. Leeuwenhoek (1718), p. 57, letter dated 29 March 1713.

201. Leeuwenhoek, *CL,* 10:57, letter dated 19 March 1694.

202. Letters dated 17 October 1687 (*CL,* 7:85–133), 9 February 1702 (Leeuwenhoek [1702], pp. 400–414), 25 December 1702 (*Phil. Trans.* 23 [1703], pp. 1304–11), and 4 November 1704 (*Phil. Trans.* 24 [1705], 1784–93).

203. Leeuwenhoek (1718), p. 67, letter dated 28 June 1713.

Chapter 3: The Substance of Living Matter

1. Pyle (1987).

2. Frank (1980), pp. 56–57; for a more detailed discussion see C. Wilson (1988) and Meinel (1988).

3. Charleton (1654), p. 115.

4. O'Brien (1975); Boas Hall (1966).

5. Power (1664), Preface, sixteenth page.

6. Ibid., p. 57.

7. Ibid., Preface, eighteenth page.

8. Ibid., Preface, eighth and ninth pages.

9. Ibid., p. 23.

10. Ibid., p. 38.

11. Ibid., p. 61.

12. Ibid., p. 82.

13. Birch (1756–57), 1:213. For surveys of the contents and genesis of *Micrographia,* see Frison (1965); A. R. Hall (1966); Harwood (1989).

14. Birch (1756–57), 3:338.

15. Ibid., 3:346, 349, 352, 358.

16. Hooke (1678), p. 305.

17. For the two lectures see Hooke (1726), pp. 257–68 and 270–73, respectively; quotation in ibid., p. 271.

18. C. Singer (1955); Frison (1965); A. R. Hall (1966).

19. T. S. Hall (1969), 1:295–311.

20. Harwood (1989).

21. Hooke (1665), p. 168.

22. Ibid., p. 176.

23. Ibid., p. 178.

24. Ibid., p. 179.

25. Ibid., p. 180.

26. Hooke (1726), p. 268.

27. Hooke (1665), p. 186.

28. Frank (1980), pp. 158–60; Guerrini (1989).

29. Hooke's manuscript of a lecture read before the Royal Society, 29 November 1693, Archives of the Royal Society CL.P.XX.84.

30. Hooke (1665), pp. 101 (charcoal), 113 (cork), 168 (feather).

31. Ibid., p. 143.

32. Ibid., p. 151.

33. Ibid., p. 210.

Chapter 4: The "Animal Oeconomy"

1. Delivered in 1665 and printed in 1669.

2. Birch (1756–57), 1:41–43; L. G. Wilson (1961).

3. Swammerdam (1637–38), 2:851: "Dat een Spier in syn contractie niet opblaast of opswelt, door de gesupposeerde invloejende en opbruisschende dierlyke geesten; maar dat een Spier in syn contractie veel eer ontswelt, of om myne gedagten beter uit te drukken, dat hy minder plaats beslaat."

4. Swammerdam (1637–38), 2:842–43: "My dan schynt niet onbillyk te volgen, dat daar niet als een een simpele en natuurlyke roering of irritatie der Senuen, tot de beweeging der Spieren nootsakelyk is, het sy dan dat die in de Hersenen, in het Merg, of ergens anders syn oorspronk neemt."

5. L. G. Wilson (1961).

6. Leeuwenhoek, *CL,* 1:111, 2:213, letters dated 1 June 1674 and 14 May 1677, respectively.

7. Birch (1756–57), 3:180.

8. Ibid., 3:397.

9. Ibid., 3:401.

10. Ibid., 3:402–3.

11. Hooke (1678), p. 311.

12. Borelli (1680–81), quoted in L. G. Wilson (1961), p. 172.

13. Theodore M. Brown (1981), p. 196.

14. Borelli (1680–81), 1:6.

15. Leeuwenhoek, *CL,* 3:385–403, letter dated (3) March 1682.

16. Ibid., 3:419–31, letter dated 4 April 1682.

17. Leeuwenhoek (1718), p. 122, letter dated 26 October 1714.

18. Ibid., pp. 208–9, letter dated 19 May 1716.

19. Ibid., p. 102, letter dated 21 August 1714: "Wanneer wy die ringwyse inkrimpingen in de vleesfibertjens gewaar werden, dat wy dan moeten vaststellen, dat de vleesmusculen, of wel yder vleesfibertje, in rust leggen; en dat wanneer de vleesmuscul beweegt, of wel sig uytrekt, dat dan de ringwyse deelen uyt de vleesfibertjens syn."

20. Leeuwenhoek, *CL,* 3:303–7, letter dated 12 November 1680.

21. Cole (1937), p. 34.

22. Birch (1756–57), 4:140.

23. Muys (1714, 1741).

24. Muys (1741), plate 1, figs. 16 and 17.

25. Muys (1714).

26. Published in a dissertation entitled *De anatome fibrarum, de motu musculorum . . . ,* which was included in his *Specimen quotuor librorum de fibra motrice et morbosa* (1702); Berg (1942); Grmek (1970a).

27. *De pulmonibus* (1661) (lung), *Epistolae anatomicae* (1665b, which contained *De cerebro* [brain] and *De lingua* [tongue] and to which was attached the originally anonymously published *De omento* [omentum]), *De viscerum structura* (1666, which contained *De renibus* [kidney], *De cerebri cortice* [cerebral cortex], and *De liene* [liver]), and *De externo tactus organo* (1665a) (skin).

28. Quoted in Hayman (1925), p. 252, from *De renibus* (1666).

29. Quoted in Adelmann (1966), 1:306, from *De cerebro* (1665b).

30. Quoted in ibid., 1:300, from *De cerebri cortice* (1666).

31. My translation of Malpighi (1697), *Riposta,* p. 139, based on a paraphrase of this paragraph in Adelmann (1966), 1:579.

32. Quoted in ibid., 1:453, from *De Structura glandularum* (1689).

33. Ibid., 1:574, who quotes from *Riposta* in *Opera posthuma.*

34. A. R. Hall and M. Boas Hall (1965–86), 4:90–91; Cavazza (1980).

35. Adelmann (1966), 2:844–45, who quotes from *De bombyce* (1669).

36. Quoted in ibid., 2:935, from *De formatione pulli in ovo* (1673).

37. Quoted in ibid., 2:957, from *De ovo incubato* (1675).

38. Quoted in ibid., from *De ovo incubato* (1675).

39. Quoted in ibid., 2:844, from *De bombyce* (1669).

40. Malpighi (1675, in 1686), pp. 29–30: "*Animalium perfectorum* structuram ab ovo emergentem continuâ auctione, & nutritione ad debitam magnitudinem deducit, agglomeratis novis particulis prioribus & jam preaexistentibus; eâ tamen ratione, ut in quocunque auctionis statu eadem primaeva forma & natura

manutenaetur, nullâ emergente de novo animalis parte praeter dentes & cornua: In *Insectis* autem, ultra augmentum, diversis dicam aetatibus, partes, quarum rudia delineamenta in juventute latitabant, emergunt, qualia sunt alae, antennae, & simil. In *Plantis*, eâdem munificentiâ, quotidianum contingit non tantum augmentum, trunco & ramis supercrescente ligneo involucro; sed à tenellis ramis novi singulo anno erumpunt surculi, quorum anticipatum inchoamentum, quasi peculiaris foetus dicitur, *gemma*, seu *oculus*."

41. For a discussion of the distinction between preexistence and preformation, see Bowler (1971).

42. Birch (1756–57), 2:498–99; Arber (1940–41).

43. Malpighi (1675, in 1686), p. 5 "Singula namque portio, quae invicem fibrarum frustula unit, cum parum interius emineat, *valvulae* vices supplet, & ita minima quaelibet guttula veluti per funem, seu per gradus, ad ingens deducitur fastigium."

44. Malpighi (1676, in 1686), p. 5.

45. Atti (1847), pp. 460–78.

46. Malpighi (1697), *Vita*, pp. 63–75.

47. Grew (1682), p. 107.

48. Ibid., p. 1.

49. Ibid., p. 20.

50. Ibid., p. 22.

51. Ibid., p. 64.

52. Ibid., p. 111.

53. Stroup (1990), see especially pp. 131–44.

54. Grew (1682), p. 118.

55. Ibid., p. 121.

56. Ibid.

57. Ibid., pp. 77–78.

58. Ibid., p. 76.

59. Ibid., p. 73.

60. Ibid., p. 1.

61. Ibid., p. 83.

62. Ibid., p. 126.

63. E.g., ibid., p. 64.

64. Ibid., pp. 239, 254.

65. Ibid., p. 133.

66. Ibid., p. 83.

67. Ibid., p. 132.

68. Ibid., p. 223.

69. Published as *Comparative Anatomy of Stomachs and Guts* as an appendix to *Musaeum regalis societatis.*

70. King (1666).

71. Leeuwenhoek, *CL*, 8:27–29.

72. Ruestow (1980); T. M. Brown (1981).

73. By Thomas Bartholin in 1653 in his *Vasa lymphatica nuper Hafniae in animantibus inventa et in homine.*

74. Frank (1980), pp. 154–63.

75. Hooke (1667).

76. Frank (1980), pp. 213–17.

77. Willis (1674–75), part 2.

78. De Heide in Willis (1681), pp. 24–25.

79. De Heide (1682), Obs. II.

80. Ibid.

81. D. L. Hall (1981), pp. 103–21.

82. Malpighi (1697), pp. 6–17.

83. Luyendijk-Elshout (1974, 1975); de Folter (1978); Ruestow (1980).

84. Craanen (1689), pp. 273–75.

85. Pitcairne (1715), p. 47.

86. Ruysch (1744), p. 698 (Dutch translation of *Museum anatomicum Ruyschianum* originally published in 1691): "Ja ik vertrouwe, dat de geheele toebereyding en scheyding der humeuren en vogten, als speekzel, melk, het zaad, &c. alleen afhangt van de hoedanigheyt, verandering en verscheide cours der uyteynden van de Vaten."

87. Bontekoe (1684), pp. 60–62.

88. Nuck (1691), p. 53 (passage between blood vessels and lymphaeducts); Cowper (1703) (passage from arteries to veins in various organs).

89. Malpighi (1666: *De renibus*) (passage from blood vessels to urinary ducts).

90. Ruysch (1744), p. 696 (Dutch translation of *Museum anatomicum Ruyschianum* originally published in 1691).

91. T. M. Brown (1981), pp. 192–237; Guerrini (1987).

Chapter 5: The Fabric of Living Beings

1. Cole (1944), p. 177.

2. Matsuo (1975); Stroup (1990).

3. Willis (1674), Preface.

4. Willis (1675), Preface (quoted in the Dutch translation of 1681): "hebben wy omtrent de Borst en de Long of de allergeheimste hoeken doorsnuffelende, alles dat de Ouden en Nieuwen geschreven, en alles dat door het ontleed-mes en door vergroot-glasen konde ontdekt werden, opgevischt en voor-gesteld."

5. Willis (1674), Preface (quoted in the Dutch translation of 1677): "hebben wy sonder hulpe of leidinge van andere Schrijvers klaar in de maag en darmen ontdekt, de zenuachtige, vleesige en de klierige rokken, ook eenige gevoel-en andere beweeg-draden, beneffens seer dichte vlechtingen van bloed-vaten, en menigte klieren."

6. Dumaitre (1964); Glick (1974); Lopez Piñero (1982).

7. A Dutch translation appeared in 1690, and an augmented English edition was edited and published by William Cowper in 1698.

8. Bidloo (1690), Preface: "Ik vertoon niets,ik zegge nogmaals niets, naar de tekening van anderen, ik haat het slaafsche werk van uittrekken."

9. Fournier (1985).

10. Hayman (1925); Grondona (1964); Belloni (1966b, 1968a); Clarke and Bearn (1968).

11. Bidloo (1715).

12. Bidloo (1685, trans. of 1690), plate 36, fig. 4.

13. *Observationes* (1673), pp. 49–51, ill.

14. Perrault (1734), plate 12.

15. Ashley Montagu (1943); Cole (1944), pp. 198–221; Williams (1976).

16. Quoted in Ashley Montagu (1943), p. 93, from the preliminary discourse to Tyson's *Phocaena* of 1680.

17. Tyson (1685).

18. Tyson (1683b).

19. Tyson (1691).

20. Tyson (1683a).

21. De Man (1905); Lindeboom (1983).

22. De Heide (1683), pp. 45–48.

23. *Observationes* (1667) and *Observationum* (1673).

24. Swammerdam (1737–38), p. 69; Lindeboom (1975c), pp. 98–99, (1981).

25. Swammerdam (1667), corollaria numbers 22 and 23.

26. Swammerdam (1669), p. 81; Lindeboom (1980) assumed that Swammerdam had learned the technique from Leeuwenhoek, citing a passage to that effect from the *Biblia naturae*, the aforementioned paragraph in Swammerdam's *Bloedeloose dierkens* having escaped his notice.

27. Lindeboom (1975c), p. 63.

28. Swammerdam (1737–38), p. 405.

29. Quoted in Lindeboom (1975c), p. 102.

30. Swammerdam (1737–38), p. 491.

31. Swammerdam (1675), p. 86: "want my dat een al te verdrietigen arbeyt docht te sijn; ende van weynigh nut."

32. Swammerdam (1737–38), p. 461.

33. Ibid., p. 2, in a paragraph that was originally published in *Bloedeloose dierkens* (1669).

34. Ibid., p. 493: "Syn soo veel Vesels te sien, als het Hoornvlies en het Oog van bovenen verdeelingen heeft: deese Vesels sluyten heel net in de holligheeden van de spherische verdeelingen van het Hoornvlies. Haar figuur van boovenen is ses hoekig en breet, in 't midden dunder, en in 't eynde spits: voorts syn sy haast altemaal van eene langte, dikte, breette ende grootte."

35. Ibid., p. 495: "Een ander of tweede soort van Vesels . . . dewelke tegens

de beschreeve Vliesen van onderen dwars aangelegt syn, ende als de fondament balken van de boven op staande pyramidale Vesels haar vertoonen. Deese Vesels verscheelen van de bovenste pyramidale, dat se in soo groote kwantiteyt niet en syn als deselve, ende ook op ver na soo subtiel niet."

36. Ibid., p. 501: "Soo syn dan deese Oogen soo gestelt, dat se de gedaantens der dingen, door een enkele voortstooting van het weeromgekaatste ligt, kunnen ontfangen."

37. Swammerdam (1675), p. 120: "Soo dat het gesight van deese beeskens, op een heel andere wijse, als in ons toegaat. Alwaar het door een vergaderingh van straalen, binnen in het oogh geschiet. Daar het selve alhier, door middel van een vergaderingh van senuachtige draatkens, toegaat; en die op de tijt als sy sien, maar eeven op haare verheventheeden, door de sienelycke hoedanigheeden en straalen van licht ende couleur geroert en beweeght worden."

38. Swammerdam (1737–38), e.g., about Hooke (p. 501) and about Malpighi (p. 410).

39. Leeuwenhoek, *CL*, 10:269–301, letter dated 10 July 1695.

40. Tyson (1685); Leeuwenhoek (1704).

41. Bonomo and Cestoni (1687).

42. Cole (1944), pp. 345–50; Peyer (1946); Boschung (1983).

43. Muralt (1682, 1683).

44. De Martin and de Martin (1970), (1983), pp. 45–47.

45. Griendel (1687), Dedication.

46. Ibid., p. 28.

47. Franceschini (1970).

48. Buonanni (1691), p. 6.

49. Ibid., pp. 39–40.

50. Franceschini (1970).

51. Sachs (1670).

52. Schrader (1681), p. 23: "Non putem tamen pulchrius & utilius ullum, etiam in Anatome fuisse inventum, quam microscopiorum, quicquid contradicant ignari hujus artificii osores."

53. Ibid., p. 23.

54. Borch (1677–79); Mentzel, "De perlis praestantissimo Muscarum genere," in *Miscellanea curiosa,* Decas II, Annus 3 (1684), 117–23; Bidloo, "De oculis et visu" in *Opera omnia.*

55. Perrault (1680), 2:337.

56. De la Hire (1678).

57. Catelan (1680, 1681a, 1681b).

58. Hooke (1665), p. 179.

59. Poupart (1700).

60. Puget (1706): "Après avoir traversé les cristallins, se va réunir dans un seul endroit de l'oeil de la mouche, pour n'y peindre qu'un seule image, & que la

convexité de toute la corneé est une condition necessaire pour causer cette réunion, en ce qu'elle dirige vers un point tous les foyers de ces cristallins."

61. See, respectively, Snelders (1982); Hooykaas (1950); Schierbeek (1950–51), chap. 3, paragraph 4.

62. Catalogus (1747); van Seters (1933).

63. Leeuwenhoek, *CL,* 8:83–93, letter dated 12 January 1689.

64. Leeuwenhoek (1702), p. 96, letter dated 9 June 1699.

65. Dobell (1932); van Seters (1933); van Cittert (1934b); Cole (1938a); Rooseboom (1939, 1950, 1959, 1967); Schierbeek (1950–51); van der Star (1953); Heniger (1973); van Zuylen (1981a, 1981b); Ford (1985); Bracegirdle (1986).

66. H. Baker (1740); van Cittert (1934b); Rooseboom (1939a, 1939b); van der Star (1953), pp. 8–9; van Zuylen (1981a, 1981b).

67. Dobell (1932); van Seters (1933); Cole (1951); Schierbeek (1950–51), chap. 2.

68. Leeuwenhoek (1704).

69. Leeuwenhoek (1718), pp. 3–4, letter dated 8 November 1712.

70. Leeuwenhoek, *CL,* 2:280, letter dated November 1677.

71. Ibid., 1:165, letter dated 7 September 1674.

72. Ibid., 2:61–161, letter dated 9 October 1676.

73. Ibid., 9:271–87, letter dated 20 December 1693; 10:269–301, letter dated 10 July 1695.

74. Ibid., 4:123–55, letter dated 7 September 1683.

75. Ibid., 3:213, letter dated 5 April 1680; see also Palm (1982).

76. Leeuwenhoek, *CL,* 1:139–49, letter dated 7 September 1674.

77. Ibid., 1:87, letter dated 24 April 1674.

78. Ibid., 4:211–41, 4:281, 5:321: letters dated, respectively, 14 April 1684, 25 July 1684, and 12 October 1685.

79. For the structure of membranes see, e.g., ibid., 2:211–13, letter dated 14 May 1677.

80. Ibid., 9:79–81, letter dated 12 August 1692.

81. Leeuwenhoek (1718), pp. 38–42, letter dated 14 March 1713.

82. Calculated in the table of weights and measures used by Leeuwenhoek, published in *CL,* 10:309.

83. Leeuwenhoek, *CL,* 1:33, letter dated 28 April 1673.

84. Swammerdam (1737–38), p. 490.

85. Leeuwenhoek, *CL,* 10:127, letter dated 30 April 1694.

86. See, e.g., Leeuwenhoek (1702), pp. 445–46, letter dated 20 April 1702; and Leeuwenhoek (1718), pp. 350–52, letter dated 6 May 1717.

87. Leeuwenhoek, *CL,* 12:227, letter dated 9 May 1698.

88. Leeuwenhoek (1702), p. 192, letter dated 2 June 1700.

89. Ford (1985); Bracegirdle (1986).

90. Leeuwenhoek (1718), pp. 168–69, letter dated 28 September 1715.

91. Ibid., p. 360, letter dated 26 May 1717.

92. Leeuwenhoek (1702), p. 306, letter dated 25 December 1700: "Moeten wy niet verbaast staan, over de onbegrypelijke veelheid, en kleynheid van deelen, waar uyt soo een staart bestaat, en wel voornamentlijk, als wy vast stellen, dat soo een dun staartjen met soo veel leden versien moet wesen, naar mate van de staarten van grootte Dieren, sal de staart sig vaardig, en aan alle kanten, konnen bewegen, en dat yder van die ledekens, niet alleen haar bysondere Musculs moeten hebben, maar ook senuwen, en aders, die het voetsel toe voeren, in 't kort, de deelen, en haar kleynheid, waar uyt de lighamen sijn te samen gestelt, sijn voor ons onbedenkelijk."

93. Leeuwenhoek, *CL*, 3:325, letter dated 12 November 1680.

94. Leeuwenhoek (1702), p. 234; Leeuwenhoek (1718), letters dated 9 June 1700 and 22 June 1714, respectively.

95. Castellani (1973).

96. Leeuwenhoek, *CL*, 1:331 and 347, letters dated 20 December 1675 and 22 January 1676.

97. Ibid., 2:115, letter dated 9 October 1676.

98. Birch (1756–57), 3:338, 346, 349, 358, 364; Hooke (1672–80), pp. 326–27.

99. Birch (1756–57), 3:352.

100. Dobell (1932), p. 148, quotes a letter by Hooke to Leeuwenhoek reporting this event.

101. Birch (1756–57), 3:366–67, 383; 4:44.

102. Hooke (1678), p. 299; Birch (1756–57), 3:366, 391, 393, 430.

103. Nicolson (1956).

104. Ploeg (1934), pp. 45–47.

105. Rooseboom (1958); Fournier (1981).

106. Huygens, *OC*, 8:96: n. 1.

107. For Picard see ibid., 19:439; for de la Hire see ibid., 22:269; for Rømer see ibid., 13 (2):704.

108. Ibid., 8:224–25: "Ik twijfel niet of de fransche nieuwsgierigheid ontrent de microscopia is nu al 't eenemael verdwenen. Wat mij aegaet, ik begin te sien, na mij dunkt, dat men 'er al langh sal moeten door sien eer men veel wijser sal worden."

109. See Cole (1926), pp. 11–14; Dobell (1932), pp. 370–72.

110. King (1693); Harris (1696); Gray (1696); Anonymous (1703a, 1703b).

111. "Philosophe" (1707).

112. Dobell (1923); Woodruff (1937); Lechevalier (1976).

113. Hooke (1726), pp. 257–68, 270–73.

114. Archives of the Royal Society CL.P.XX.84, manuscript of Hooke's lecture read before the Royal Society on 29 November 1693.

115. In a passage quoted, among many others, by Farley (1972), p. 17, from Nicolas Malebranche's *De la recherche de la vérité.*

116. Huygens, *OC,* 13 (2), pp. 698–732.
117. Beyerinck (1913); Smit and Heniger (1975).
118. Leeuwenhoek, *CL,* 3:333, letter dated 12 November 1680.
119. Huygens, *OC,* 13 (2), p. 731: "Eau de poivre vieille beaucoup de H, plusieurs N, 2 ou 3 M a queue. quantitè de tres petits comme points noirs qui remuoient beaucoup, comme hier, peu de K."
120. Point made by van Berkel (1982); examples are Dobell (1932) and Schierbeek (1950–51).
121. Huygens, *OC,* 13 (2), p. 717: "Ainsi tous les mesmes animaux sont venus dans cette eau, que dans celle de M. Romer du mesmes poivre, mais plus tard dans la miene parce qu'elle avoir estè mise plus tard à infuser."
122. Dobell (1932), p. 371.
123. Anonymous (1703a), p. 1366.
124. Huygens, *OC,* 13 (2), p. 727: "Devant la quelle parfois il y avoit une sorte circulation de brins d'ordure."
125. Leeuwenhoek (1718), p. 6, letter dated 28 June 1713.
126. Huygens, *OC,* 13 (2), p. 710: "sinon qu'ils ont la facultè d'enfler une partie de leur corps et cela successivement depuis la teste jusqu' à la queue, de sorte que cette enflure ou bosse parcourt fort subitement toute la longueur du corps et puis recommence depuis la teste ce qui suffit pour faire avancer l'animal."
127. Leeuwenhoek, *CL,* 2:145, letter dated 9 October 1676.
128. Joblot (1718), 2:15.
129. King (1693).
130. Huygens, *OC,* 13 (2), pp. 716–17.
131. Ibid., 13 (2), p. 719.
132. Ibid., 13 (2), p. 730.
133. Leeuwenhoek, *CL,* 3:261–67, letter dated 14 June 1680.
134. "Philosophe" (1707), p. 11: "Mais ne se battroient-ils jamais que deux à deux?"
135. Huygens, *OC,* 13 (2), p. 726.
136. Anonymous (1703a), p. 1366. See also Hooke (1678); Harris (1696); Joblot (1718).
137. Ibid., p. 1367.
138. Huygens, *OC,* 13 (2), p. 726.

Chapter 6: Five Heroes of Microscopic Science

1. Lindeboom (1981), pp. 107–8; based on a passage in Leeuwenhoek, *CL,* 1:143–45, letter dated 7 September 1674.
2. Adelmann (1975), 2, letters 373, 390.
3. Ibid., 2, letters 378, 382.
4. Adelmann (1966), 2: 714.

5. Adelmann (1975), 2, letters 372, 373, 390, 395, 397.

6. Leeuwenhoek, *CL*, 2, letters 37, 38, 39, 40, 42; 3; letters 43, 48, 51, 53, 54, 55, 56, 57, 59, 61, 63, 65, 66, 67, 68.

7. Ibid., 2, letters 38 and 39.

8. Ibid., 2:5–13, letter dated 21 April 1676.

9. Ibid., 2:25–39.

10. Birch (1756–57), 2:476–80.

11. De Clercq (1991).

12. Quoted in Lindeboom (1975c), p. 64; letter written ca. 1671.

13. Malpighi (1697), pp. 59–60.

14. Swammerdam (1737–38), pp. 315–16.

15. Malpighi (1697), pp. 60–62, e.g., plate 3.

Chapter 7: Measuring the Impact of the Microscope

1. Webster (1967).

2. See, e.g., Nicolson (1956), p. 26, for Samuel Pepys's reaction; Huygens, *OC*, 5:277, 282, 320; and review in *Philosophical Transactions* 1 (1665–66), pp. 27–32.

3. Archives of the Royal Society CL.P.XX 84.

4. Hooke (1665), Preface, fourth page.

5. P. Lyonet, *Traité anatomique de la chenille.*

6. O. F. Müller, *Animalcula infusoria.*

7. J. Hedwig, *Theoria generationis et fructificationis plantarum cryptogamicarum.*

8. King (1965).

9. Réaumur (1734–42, 6 vols.), see, respectively, vol. 2, plate 5, and vol. 5, plate 32.

10. Archives of the Royal Society CL.P.XX 84.

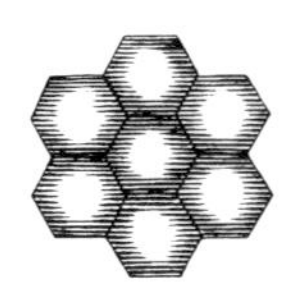

Bibliography

Primary Materials

Adams, George (1746). *Micrographia illustrata, or, the knowledge of the microscope explain'd.* London: Printed for, and sold by the author, 1746.

Anonymous (1703a). "An extract of some letters sent to Sir C.H. relating to some microscopal observations. Communicated by Sir C.H. to the publisher." *Philosophical Transactions* 23 (1702–3), pp. 1357–72 (=1361–76).

Anonymous (1703b). "Two letters from a gentleman in the country, relating to Mr. Leuwenhoeck's letter in Transaction, no. 282. Communicated by Mr. C." *Philosophical Transactions* 23 (1702–3), pp. 1494–1501.

Baker, Henry (1740). "An account of Mr. Leeuwenhoek's microscopes." *Philosophical Transactions* 41–42 (1740–41), pp. 503–19.

———. (1742). *The microscope made easy* . . . London: Printed for R. Dodsley, 1742.

Balthasar, Theodor (1710). *Micrometria, hoc est, de micrometrorum, tubis opticis* . . . Basel: Christian Erlange, 1710.

Beeckman, Isaac (1604–34). *Journal tenu par Isaac Beeckman de 1604 à 1634, Publié avec une introduction et des notes par C. de Waard.* 4 vols. The Hague: Nijhoff, 1939–53.

Bidloo, Goverd (1690). *Ontleding des menschelyken lichaams.* Amsterdam: de weduwe van Joannes van Someren, 1690. Translation of *Anatomia corporis humani,* Amsterdam, 1685.

———. (1715). *De oculis et visu variorum animalium observationes physicoanatomicae.* Leiden, 1715. Also in Bidloo's *Opera omnia,* Leiden, 1715.

Birch, Thomas (1756–57). *The history of the Royal Society of London for improving of natural knowledge, from its first rise* . . . 4 vols. London: Printed for A. Millar, 1756–57. Reprint, Hildesheim: Georg Olms, 1968.

Boerhaave, Herman (1737). "Vita D. Joannis Swammerdammii. Het leven van den heer Jan Swammerdam," introduction to Jan Swammerdam, *Biblia naturae.* Leiden: Isaak Severinus, Boudewyn van der Aa, Pieter van der Aa, 1737, pp. 15–49.

Bonomo, Giovan Cosimo, and Giacinto Cestoni (1687). *Epistola che contiene osservazioni intorno a'pellicelli del corpo umano.* Firenze: P. Martini, 1687.

Bontekoe, Cornelis (1684). *Korte verhandeling van 's Menschen Leven, Gesondheid, Siekte, en Dood* . . . The Hague: Peter Hagen, 1684.

Borch, Ole (1660–65). *Olai Borrichii itinerarium 1660–1665: The journal of the Danish polyhistor Ole Borch.* 4 vols. Copenhagen: Danish Society of Language and Literature, 1983.

———. (1677–79). "Aegus Islandicus." In *Acta medica et philosophica Hafniensa* 5 (1677–79), 218–22.

Borel, Pierre (1656). *Observationum microscopicarum centuria.* The Hague: Ex officina A. Vlacq, 1656.

Borelli, Giovanni Alfonso (1680–81). *De motu animalium . . . opum posthumum.* 2 vols. Rome: ex Typographia Angeli Bernabo, 1680–81.

Buonanni, Philippo (1691). *Micrographia curiosa sive rerum minutissimarum observationes* . . . Rome: Typis Dominici Antonii Herculis, 1691.

Catalogus (1747). *Catalogus rarissimorum et exquisitissimorum microscopiorum . . . invenit et fabrefecit celeberrimus Anthonius a Leeuwenhoek . . . Catalogus van het vermaarde cabinet van vergrootglasen . . . nagelaten door wijlen den heer Anthony van Leeuwenhoek* . . . Delft: Boitet, 1747.

Catelan, Abbé de (1680). "Observations touchant les deux parties des insectes qu'on prend d'ordinaire pour leurs yeux." *Journal des sçavans* (1680), pp. 167–70.

———. (1681a). "Autres observations touchant les yeux des insectes." *Journal des sçavans* (1681), pp. 92–94.

———. (1681b). "Suite des observations de M. l'Abbé de Catelan touchant les yeux des insectes." *Journal des sçavans* (1681), pp. 140–42.

Charleton, Walter (1654). *Physiologia Epicuro-Gassendo-Charltoniana, or a fabrick of science natural, upon the hypothesis of atoms, founded by Epicurus, repaired by Petrus Gassendus, augmented by Walter Charleton.* London: Printed by Thomas Newcomb, 1654. Reprinted, Sources of Science 31. New York: Johnson Reprint Co., 1966.

Chérubin d'Orleans, Père (1677). *La vision parfaite, ou le concours des deux axes de la vision en un seul point de l'objet.* Paris: Sebastien Mabre–Cramoisy, 1677.

Cowper, William (1703). "An account of divers schemes of arteries and veins . . . : with some chirurgical observations, and figures after the life." *Philosophical Transactions* 23 (1702–3), pp. 1177–1201.

Craanen, Theodor (1689). *Tractatus physico-medicus de homine.* Leiden: apud Petrum van der Aa, 1689.

Divini, Eustachio (1668). "Another extract out of the Italian Journal [Giornale de Letterati, 1667], being a description of a microscope of a new fashion, by the means whereof there hath been seen an animal lesser than any of those seen hitherto." *Philosophical Transactions* 3 (1668), pp. 842.

Folkes, Martin (1723). "Some account of Mr. Leeuwenhoek's curious microscopes, lately presented to the Royal Society." *Philosophical Transactions* 32 (1722–23), pp. 446–53.

Fontana, Francesco (1646). *Novae coelestium, terrestriumque rerum observationes.* Naples: Cassarum, 1646.

Gray, Steven (1696). "Several microscopical observations and experiments." *Philosophical Transactions* 19 (1695–97), pp. 280–87.

Grew, Nehemiah (1672). *The anatomy of vegetables begun. With a general account of vegetation founded thereon.* London, 1672.

———. (1681). *Musaeum Regalis Societatis . . . Whereunto is Subjoyned the Comparative Anatomy of Stomachs and Guts.* London: Printed for the author, 1681.

———. (1682). *The anatomy of plants. With an idea of a philosophical history of plants.* London: Rawlins, 1682.

———. (1701). *Cosmologia sacra or a discourse of the universe as it is the creature and kingdom of God . . .* London, 1701.

Griendel von Ach, Johann Franz (1687). *Micrographia nova, oder neucurieuse Beschreibung verschiedener kleiner Körper . . .* Nuremberg: J. Ziegeri, 1687.

Harris, John (1696). "Some microscopical observations of vast numbers of animalcula seen in water." *Philosophical Transactions* 19 (1695–97), pp. 254–59.

Hartsoeker, Nicolaas (1730). *Cours de physique. Accompagné de plusieurs piéces concernant la physique qui ont déja paru, et d'un extrait critique des lettres de M. Leeuwenhoek.* The Hague: Jean Swart, 1730.

Heide, Anton de (1682). *Nieuw ligt der apothekers . . . Beneffens eenige ontleed- genees- en heelkundige waarnemingen . . .* Amsterdam: d'Erfgenamen van Joannes Janssonius van Waasberge, 1682.

———. (1683). *Anatome mytuli, Belgice Mossel, structuram elegantem ejus motumque mirandum exponens . . .* Amsterdam: apud Janssonio-Waesbergios, 1683.

Hevelius, Johannes (1673–79). *Machinae coelestis . . .* 3 vols. Danzig: Simon Reiniger, 1673–79.

Hire, Philippe de la (1678). "Nouvelle découverte des yeux de la mouche et des autres insectes volans, faite à la faveur du microscope" *Journal des sçavans* (1678), pp. 187–88.

Hooke, Robert (1665). *Micrographia: or some physiological descriptions of minute bodies made by magnifying glasses. With observations and inquiries thereupon.* London: Jo. Martyn, 1665.

———. (1667). "An account of an experiment made by M. Hooke, of preserving animals alive by blowing through their lungs with bellows." *Philosophical Transactions* 2 (1667), pp. 539–40.

———. (1672–80). *The diary of Robert Hooke M.A., M.D., F.R.S. 1672–1680. Transcribed from the original in the possession of the Corporation of the City of London (Guildhall Library).* Ed. Henry W. Robinson and Walter Adams. London: Taylor and Francis, 1935.

———. (1678). "Microscopium: or, some new discoveries made with and concerning microscopes." In *Lectures and collections made by Robert Hooke. Cometa . . . microscopium.* London: Printed for J. Martyn, 1678, pp. 81–104. Reprinted in R. T. Gunther, *Early science in Oxford,* vol. 8. Oxford, 1931, pp. 297–320.

———. (1705). *The posthumous works of Robert Hooke containing his Cutlerian lectures, and other discourses, . . . Publish'd by Richard Waller.* London: Royal Society, 1705. Reprint, London: F. Cass, 1971.

———. (1726). *Philosophical experiments and observations.* Published by W. Derham. London: Royal Society, 1726. Reprint, London: Cass, 1967.

Huygens, Christiaan (1888–1950). *Oeuvres complètes de Christiaan Huygens.* Publiées par la Société Hollandaises des Sciences. 22 vols. The Hague: Nijhoff, 1888–1950.

Joblot, Louis (1718). *Descriptions et usages de plusieurs nouveaux microscopes, tant simple que composez . . .* Paris: chez J. Collombat, 1718.

Jurin, James (1718). "De motu aquarum fluentium." *Philosophical Transactions* 30 (1717–19), pp. 748–66.

Kerckring, Theodorus (1670). *Spicilegium anatomicum . . .* Amsterdam: sumptibus Andreae Frisii, 1670.

King, Edmund (1666). "Some considerations concerning the parenchymous parts of the body." *Philosophical Transactions* 1 (1665–66), pp. 316–20.

———. (1693). "Several observations and experiments on the animalcula, in pepper-water, etc." *Philosophical Transactions* 17 (1693), pp. 861–65.

Kircher, Athanasius (1646). *Ars magna lucis et umbrae in decem libros digesta . . .* Rome: Sumptibus Hermanni Scheus, ex Typographia Ludovici Grignani, 1646.

Leeuwenhoek, Antoni van (1693). *Derde vervolg der brieven . . .* Delft: Henrik van Krooneveld, 1693.

———. (1702). *Sevende vervolg der brieven . . .* Delft: Henrik van Krooneveld, 1702.

———. (1704). "Concerning Cochineel." *Philosophical Transactions* 24 (1704–5), pp. 1614–28.

———. (1718). *Send-brieven . . .* Delft: Adriaan Beman, 1718.

———. (1721). "Upon the membranes enclosing the fasciculi of fibres." *Philosophical Transactions* 31 (1720–21), pp. 129–34.

———. (1939–). *Alle de brieven van Antoni van Leeuwenhoek. The collected letters of Antoni van Leeuwenhoek.* Amsterdam: Swets and Zeitlinger, 1939–.

Malpighi, Marcello (1661). *De pulmonibus observationes anatomicae.* Bononiae: Typis Io. Baptistae Ferronii, 1661. Reprinted in Malpighi's *Opera Omnia.* London, 1686.

———. (1665a). *De externo tactus organo anatomica observatio.* Neapoli: apud Aegidium Longum, 1665. Reprinted in Malpighi's *Opera Omnia.* London, 1686.

———. (1665b). *Epistolae anatomicae de cerebro, ac lingua . . .* Bologna: Typis Antonii Pisarii, 1665. Reprinted in Malpighi's *Opera Omnia.* London, 1686.

———. (1666). *De viscerum structura exercitatio anatomica.* Bologna: Ex typographia Iacobi Montii, 1666. Reprinted in Malpighi's *Opera Omnia.* London, 1686.

———. (1669). *Dissertatio epistolica de bombice . . .* London: apud Joannem Martyn et Jacobum Allestry, 1669. Reprinted in Malpighi's *Opera Omnia.* London, 1686.

———. (1673). *Dissertatio epistolica de formatione pulli in ovo . . .* London: apud Joannem Martyn, 1673. Reprinted in Malpighi's *Opera Omnia.* London, 1686.

———. (1675–79). *Anatome plantarum. Cui subjungitur appendix . . . De ovo incubato observationes . . .* London: Impressis Johannis Martyn, 1675–79. Reprinted in Malpighi's *Opera Omnia.* London, 1686.

———. (1686). *Opera omnia.* London: apud Robertum Scott et Georgium Wells, 1686.

———. (1697). *Opera posthuma . . . Quibus praefixa est ejusdem vita à seipso scripta.* London: Impensis A. et J. Churchill, 1697.

Mentzel, Christian (1684). "De perlis praestantissimo Muscarum genere." *Miscellanea curiosa sive ephemeridum medico-physicarum Germanicarum Academiae Naturae Curiosorum.* Decuriae II, Annus Tertius, Anni 1684, pp. 117–23.

Muralt, Johann von (1682). "Examen anatomicum Grylli sylvestris." In *Miscellanea curiosa sive ephemeridum medico-physicarum Germanicarum Academiae Naturae Curiosorum.* Decuriae II, Annus Primus, Anni 1682, pp. 142–47.

———. (1683). "De Gryllo-talpa." In *Miscellanea curiosa sive ephemeridum medico-physicarum Germanicarum Academiae Naturae Curiosorum,* Decuriae II, Annus secundus, Anni 1683, pp. 58–59.

Muys, Weyer Willem (1714). "An account of several observations concerning the frame and texture of the muscles." *Philosophical Transactions* 29 (1714–16), pp. 59–61.

———. (1741). *Investigatio fabricae, quae in partibus musculis componentibus extat.* Leiden: apud Joh. Langerak, 1741.

Nuck, Anton (1691). *Defensio ductuum aquosorum, cui accedunt, solutionum apologeticarum eversiones.* Leiden: apud Jordanum Luchtmans, 1691.

Observationum (1667, 1673), *Observationes anatomicae selectiores pars prima*. Amsterdam: Caspar Commelin, 1667. *Observationum anatomicarum collegii privati Amstelodamensis pars altera*. Amsterdam: apud Casparum Commelinum, 1673.

Odierna, Giovanbattista (1644). *L'occhio della mosca*. Palermo: per Decio Cirillo, 1644.

Oldenburg, Henry (1965–86). *The correspondence of Henry Oldenburg*. Ed. and trans. A. R. Hall and M. B. Hall. 13 vols. Madison: University of Wisconsin Press, 1965–86.

Perrault, Claude (1680). *Essais de physique, ou recueil de plusieurs traités touchant les choses naturelles*. 3 vols. Paris: Chez Jean Baptiste Coignard, 1680.

———. (1734). *Memoires pour servir à l'histoire naturelle des animaux*. Paris: Gabriel Martin, 1734.

"Philosophe" (1707). "Un philosophe, ami de M.Carré." *Histoire de l'Académie Royale des Sciences* (Année 1707), pp. 10–11.

Pitcairne, Archibald (1715). *The whole works of Dr. Archibald Pitcairne, published by himself*. London: E. Curll, 1715.

Poupart, François (1700). "Part of Monsieur Poupart's letter to Dr. Martin Lister, F.R.S. concerning the insect called Libella." *Philosophical Transactions* 22 (1700–1701), pp. 673–76.

Power, Henry (1664). *Experimental philosophy, in three books: containing new experiments, microscopical, mercurial, magnetical*. London: Printed by T. Roycroft, for John Martin, 1664. Reprint, with the addition of Power's notes, corrections and emendations and a new introduction by Marie Boas Hall. New York: Johnson Reprint Co., 1966.

Puget, Louis (1706). *Observations sur la structure des yeux de divers insectes*. Lyons, 1706.

Ruysch, Frederik (1744). *Alle de ontleed- genees- en heelkundige werken*. 3 vols. Amsterdam: Janssoons van Waesberge, 1744.

Sachs, Philip Jacob (1670). "Messis observationum microscopicarum è variis authoribus collectarum." *Miscellanea curiosa* 1 (1670), pp. 34–49.

Schrader, Friedrich (1681). *Dissertatio epistolica de microscopiorum usu in naturali scientia et anatome*. Göttingen: typis Johannis Christophori Hampii, 1681.

Stelluti, Francesco (1630). *Persio tradotto in verso sciolto e dichiarato*. Rome: Giacomo Mascardi, 1630.

Stensen, Niels (1664). *De musculis et glandulis observationum specimen*. Hafniae: lit. M. Godiechenii, 1664.

Swammerdam, Jan (1667). *Tractatus physico-anatomico-medicus de respiratione usuque pulmonum*. Leiden: Gaasbeeck, 1667. Reprinted in *Opuscula selecta Neerlandicorum de arte medica*. Amsterdam, 1927, vol. 6, pp. 46–181.

———. (1669). *Historia generalis ofte algemeene verhandeling der bloedeloose dierkens.* Utrecht: van Dreunen, 1669.

———. (1675). *Ephemeri vita. Of afbeeldingh van 's menschen leven, . . .* Amsterdam: Abraham Wolfgang, 1675.

———. (1681). "Le cabinet de Mr. Swammerdam, docteur en medecine, ou catalogue de toutes sortes d'insectes." In Melchisedec Thévenot, *Receuil des voyages.* Paris, 1681.

———. (1737–38). *Biblia naturae; sive historia insectorum, . . . Hier bij komt een voorreeden . . . Herman Boerhaave. De Latynsche overzetting heeft bezorgt Hieronimus David Gaubius.* 2 vols. Leiden: Isaak Severinus, 1737–38.

Tyson, Edward (1683a). "Lumbricus latus." *Philosophical Transactions* 13 (1683), pp. 113–44.

———. (1683b). "Lumbricus teres, or some anatomical observations on the Round Worm bred in human bodies." *Philosophical Transactions* 13 (1683), pp. 154–61.

———. (1685). (Drawings of cochinal fly). *Philosophical Transactions* 15 (1685), p. 1202, drawing facing p. 1186.

———. (1691). "Lumbricus hydropicus; or An essay to prove that Hydatides often met with in morbid animal bodies, are a species of worms, or imperfect animals." *Philosophical Transactions* 16 (1686–87), pp. 506–10.

Verduc, Jean Baptiste (1682). *Discours sur l'utilité du microscope, dans les decouvertes d'anatomie, de physique et de medecine.* Paris, 1682.

Willis, Thomas (1664). *Cerebri anatome: Cui accessit nervorum descriptio et usus.* London: J. Flesher, 1664.

———. (1674–75). *Pharmaceutice rationalis: Sive diatriba de medicamentorum operationibus in humano corpore . . .* Oxford: E Theatro Sheldoniano, Prostant apud Robertum Scott, 1674. Pars secunda, Oxford, 1675.

———. (1677). *d'Algemeene en byzondere wercking der genees- middelen in 's menschen lichaam. Uyt het Lat. vert. en verrijckt door A. d'Heide.* Middelburg, 1677.

———. (1681). *Vervolg of tweede deel der redenkundige verhandeling van de kragt en werking der genees-middelen. Uit het Lat.vert. en met aantek. verrijkt door A. D[e] H[eide].* Amsterdam: By Wilh. Goeree, 1681.

Secondary Sources

Abetti, Giorgio (1974). "Odierna (or Hodierna), Gioanbatista." In *Dictionary of scientific biography.* Ed. Charles G. Gillespie. New York: Scribner's, 1974, vol. 10, p. 176.

Adelmann, Howard B. (1966). *Marcello Malpighi and the evolution of embryology.* 5 vols. Ithaca: Cornell University Press, 1966.

———. (1975). *Marcello Malpighi: The correspondence of Marcello Malpighi.* 5 vols. Ithaca: Cornell University Press, 1975.

Arber, Agnes (1940–41). "Nehemiah Grew and Marcello Malpighi." *Proceedings of the Linnean Society,* 153d session (1940–41), pp. 218–37.

———. (1942). "Nehemiah Grew (1641–1712) and Marcello Malpighi (1628–1694): an essay in comparison." *Isis* 34 (1942), pp. 7–16.

Atti, Gaetano (1847). *Notizie edite ed inedite della vita e delle opere di Marcello Malpighi e di Lorenzo Bellini.* Bologna: Tipografia Governativa alla Volpe, 1847.

Baas, Pieter (1982). "Leeuwenhoek's contributions to wood anatomy and his ideas on sap transport in plants." In *Antoni van Leeuwenhoek 1632–1723.* Ed. L. C. Palm and H. A. M. Snelders. Amsterdam: Rodopi, 1982, pp. 79–107.

Baker, John R. (1945). "The discovery of the uses of colouring agents in biological micro-technique." *Journal of the Quekett Microscopical Club,* series 4, 1 (1943) no. 6.

———. (1952). *Abraham Trembley of Geneva: Scientist and philosopher, 1710–1784.* London: Arnold, 1952.

Balaguer Perigüell, Emilio (1971). "La introduccion de la metodologia moderna en biologia: El De Motu Animalium de J. A. Borelli (1608–1679)." *Episteme* 5 (1971), pp. 243–62.

Bastholm, E. (1950). *The history of muscle physiology from the natural philosophers to Albrecht von Haller: A study of the history of medicine with a summary in Danish.* Acta Historica scientiarum et medicinalium, vol. 7. Copenhagen: Munksgaard, 1950.

Bäumer, Äenne (1987). "Zur Verhältnis von Religion und Zoologie in 17. Jahrhundert (William Harvey, Nathaniel Highmore, Jan Swammerdam)." *Berichte zur Wissenschaftsgeschichte* 10 (1987), pp. 69–81.

Becking, L. B. (1924). "Antoni van Leeuwenhoek, immortal dilettant (1632–1723)." *Scientific Monthly* 18 (1924), pp. 547–54.

Bedini, Silvio A. (1963). "Seventeenth century Italian compound microscopes." *Physis* 5 (1963), pp. 383–422.

Belloni, Luigi (1962). "Micrografia illusoria e 'animalcula.'" *Physis* 4 (1962), pp. 65–73.

———. (1963). "I capillari sanguigni nelle Tavole del Malpighi." *Physis* 5 (1963), pp. 70–77.

———. (1964). "Essais d'anatomie de texture au XVIe siècle." In *Aktuelle Probleme aus der Geschichte der Medizin: Verhandlungen des 19. Internationalen Kongresses für Geschichte der Medizin, Basel, September 1964.* Basel: Karger, 1966, pp. 7–11.

———. (1965a). "Die Entstehungsgeschichte der mikroskopischen Anatomie." *Medizinische Monatschrift* (1965), pp. 113–22.

———. (1965b). "Zur Geschichte der tierforschenden Mikroskopie." *Nova Acta Leopoldina N.F.* 30 (1965), pp. 443–58.

———. (1965c). "I trattati di M. Malpighi sulla struttura della lingua e della cute." *Physis* 7 (1965), pp. 431–75.

———. (1965d). "Die Eroberung des Nephrons." *Berliner Medizin* 16 (1965), pp. 92–100.

———. (1966a). "Auf dem Wege zur Elementardrüsse als Sekretionsmaschine: Forschungen des Kreises um Borelli (Auberius, Bellini-Zambucari, Malpighi)." In *Medizingeschichte im Spektrum.* Sudhoffs Archiv Beiheft 7. Wiesbaden: Steiner, 1966, pp. 11–29.

———. (1966b). "La neuroanatomia di Marcello Malpighi." *Physis* 8 (1966), pp. 253–66.

———. (1967). "I primi passi della microscopia ad opera di Galileo e della sua Scuola." In *Atti del Symposium Internazionale "Galileo Galilei nella Storia e nella Filosofia della Scienza" (Firenze-Pisa 14–16 settembre 1964).* Firenze: 1967, pp. 129–34.

———. (1968a). "Die Neuroanatomie von Marcello Malpighi." In *Steno and brain research.* Ed. G. Scherz. Oxford: Pergamon Press, 1968, pp. 193–206.

———. (1968b). "Stensen-Andenken in Italien." In *Steno and brain research.* Ed. G. Scherz. Oxford: Pergamon Press, 1968, pp. 171–80.

———. (1969). "Il primo ventennio della microscopia (Galilei 1610—Harvey 1628): Dalla microscopia alla anatomia microscopica dell'insetto." *Clio medica* 4 (1969), pp. 179–90.

———. (1971). "De la théorie atomistico-mecaniste à l'anatomie subtile (de Borelli à Malpighi à Morgagni) et de l'anatomie subtile à l'anatomie pathologique (de Malpighi à Morgagni)." *Clio medica* 6 (1971), pp. 99–107.

———. (1974). "Malpighi, Marcello." In *Dictionary of scientific biography.* Ed. Charles G. Gillispie. New York: Scribner's, 1974, vol. 9, pp. 62–66.

———. (1975). "Marcello Malpighi and the founding of anatomical microscopy." In *Reason, experiment and mysticism in the scientific revolution.* Ed. M. L. Righini Bonelli and William R. Shea. New York: Science History Publications, 1975, pp. 95–110.

———. (1985). "Athanasius Kircher: Seine Mikroskopie, die Animalcula und die Pestwürmer." *Medizinhistorisches Journal* 20 (1985), pp. 58–65.

Bennett, James A. (1980). "Robert Hooke as mechanic and natural philosopher." *Notes and Records of the Royal Society of London* 35 (1980), pp. 33–48.

———. (1989). "Hooke's instruments for astronomy and navigation." In *Robert Hooke: New studies.* Ed. Michael Hunter and Simon Schaffer. Woodbridge: Boydell Press, 1989, pp. 21–32.

Berg, Alexander (1942). "Die Lehre von der Faser als Form-und Funktionselement des Organismus: Die Geschichte des biologisch-medizinischen Grundproblems vom Kleinsten Bauelement des Körpers bis zur Begründung der Zellenlehre." *Virchows Archiv für pathologische Anatomie und Physiologie und für klinische Medizin* 309 (1942), pp. 333–460.

Berkel, Klaas van (1982). "Intellectuals against Leeuwenhoek." In *Antoni van Leeuwenhoek 1632–1723: Studies on the life and work.* Ed. L. C. Palm and H. A. M. Snelders. Amsterdam: Rodopi, 1982, pp. 187–209.

Bessis, Marcel, and G. Delpech (1981). "Discovery of the red blood cell with notes on priorities and credits of discoveries, past, present and future." *Blood Cells* 7 (1981), pp. 447–80.

Beyerinck, Martinus W. (1913). "De infusies en de ontdekking der bakteriën." In *Jaarboek der Koninklijke Akademie van Wetenschappen 1913.* Amsterdam: North–Holland.

Bodemer, Charles W. (1973). "The microscope in early embryological investigation." *Gynecologic Investigation* 4 (1973), pp. 188–209.

Bolam, Jeanne (1973). "The botanical works of Nehemiah Grew F.R.S. (1641–1712)." *Notes and Records of the Royal Society* 27 (1973), pp. 219–31.

Boschung, Urs (1983). *Johannes von Muralt, 1645–1733: Arzt, Chirurg, Anatom, Naturforscher, Philosoph.* Zurich: Hans Rohr, 1983.

Bowler, Peter J. (1971). "Preformation and pre–existence in the seventeenth century: A brief analysis." *Journal of the History of Biology* 4 (1971), pp. 221–44.

Bracegirdle, Brian (1978a). *A history of microtechnique: The evolution of the microtome and the development of tissue preparation.* London: Heinemann, 1978.

———. (1978b). "The performance of seventeenth- and eighteenth-century microscopes." *Medical History* 22 (1978), pp. 187–95.

———. (1984). "Techniques of specimen-preparation at the time of Leeuwenhoek." In *Beads of glass: Leeuwenhoek and the early microscope. Catalogue of an exhibition.* Ed. B. Bracegirdle. London: Science Museum, 1984.

———. (1986). "Famous microscopists: Antoni van Leeuwenhoek, 1632–1723." *Proceedings of the Royal Microscopical Society* 21 (1986), pp. 367–73.

Bradbury, Saville (1967a). *The evolution of the microscope.* Oxford: Pergamon Press, 1967.

———. (1967b). "The quality of the image produced by the compound microscope: 1700–1840." In *Historical aspects of microscopy: Papers read at a one-day conference held by the Royal Microscopical Society at Oxford, 18 March, 1966.* Ed. S. Bradbury and G. l'E. Turner. Cambridge: Heffer, 1967, pp. 151–73.

Brown, Harcourt (1974). "Peiresc, Nicolas Claude Fabri de." In *Dictionary of scientific biography.* Ed. Charles G. Gillispie. New York: Scribner's, 1974, vol. 10, pp. 488–92.

Brown, Theodore M. (1971a). "Descartes, René du Perron." In *Dictionary of scientific biography.* Ed. Charles G. Gillispie. New York: Scribner's, 1971, vol. 4, pp. 51–65.

———. (1971b). *The posthumous works of Robert Hooke.* London: Cass, 1971.

———. (1977). "Physiology and the mechanical philosophy in mid-seventeenth century England." *Bulletin of the History of Medicine* 51 (1977), pp. 25–54.

———. (1981). *The mechanical philosophy and the "animal oeconomy."* New York: Arno Press, 1981. Reprint of "The mechanical philosophy and the 'animal oeconomy': A study in the development of English physiology in the seventeenth and early eighteenth century." Thesis, Princeton University, 1968.

Bynum, William F. (1973). "The anatomical method, natural theology, and the functions of the brain." *Isis* 64 (1973), pp. 445–68.

Carré, Marie-Rose (1974). "A man between two worlds: Pierre Borel and his Discours nouveau prouvant la pluralités des mondes of 1657." *Isis* 65 (1974), pp. 322–35.

Carruthers, William (1902). "The president's address: On the life and work of Nehemiah Grew." *Journal of the Royal Microscopical Society* (1902), pp. 129–41.

Castellani, Carlo (1973). "Spermatozoan biology from Leeuwenhoek to Spallanzani." *Journal of the History of Biology* 6 (1973), pp. 37–68.

Catalogue (1929). *Catalogue of the printed books and pamphlets in the library of the Royal Microscopical Society.* London: Royal Microscopical Society, 1929.

Cavazza, Marta (1980). "Bologna and the Royal Society in the seventeenth century." *Notes and Records of the Royal Society of London* 35 (1980), pp. 105–23.

Chabbert, Pierre (1968). "Pierre Borel (1620?–1671)." *Revue d'histoire des sciences* 21 (1968), pp. 303–43.

Charante, Nicolaus H. van (1844). *Dissertatio historico-medica inauguralis de Antonii Leeuwenhoeckii meritis in quasdam partes anatomiae microscopicae.* Leiden: Gebhard, 1844.

Cittert, Pieter H. van (1934a). *Descriptive catalogue of the collection of microscopes in charge of the Utrecht University Museum with an introductory historical survey of the resolving power of the microscope.* Groningen: Noordhoff, 1934.

———. (1934b). "The optical properties of the 'Van Leeuwenhoek' microscope in possession of the University of Utrecht." *Proceedings Koninklijke Akademie Amsterdam* 337 (1934), pp. 290–93.

Clarke, Edwin, and J. G. Bearn (1968). "The brain 'glands' of Malpighi elucidated by practical history." *Journal of the History of Medicine and Allied Sciences* 23 (1968), pp. 309–30.

Clay, Reginald S., and Thomas H. Court (1932). *The history of the microscope: Compiled from original instruments and documents, up to the introduction of the achromatic microscope.* London: Griffin, 1932. Reprint, London: Holland Press, 1985.

Clercq, Peter de (1991). "Exporting scientific instruments around 1700: The Musschenbroek documents in Marburg." *Tractrix* 3 (1991), pp. 79–120.

Cole, Francis J. (1921). "The history of anatomical injections." In *Studies in the history and method of science.* Ed. Charles Singer. Oxford: Clarendon Press, 1921, pp. 285–343.

———. (1926). *The history of protozoology: Two lectures delivered before the University of London at King's College in May 1925.* London: University of London Press, 1926.

———. (1930). *Early theories of sexual generation.* Oxford: Clarendon Press, 1930.

———. (1937). "Leeuwenhoek's zoological researches." *Annals of Science* 2 (1937), pp. 1–46, 185–235.

———. (1938a). "Microscopic science in Holland in the seventeenth century." *Journal of the Quekett Microscopical Club,* series 4/1 (1938).

———. (1938b). *Observationes anatomicae selectiores Amstelodamensium 1667–1673.* Berkshire: University of Reading, 1938.

———. (1944). *A history of comparative anatomy: From Aristotle to the eighteenth century.* 2d ed. London, Macmillan, 1949.

———. (1951). "History of micro-dissection." *Proceedings of the Royal Society B,* 138 (1951), pp. 159–87.

Cole, Francis J., and Nellie B. Eales (1917). "The history of comparative anatomy: I. A statistical analysis of the literature." *Science Progress* 11 (1916–17), pp. 578–96.

Daumas, Maurice (1972). *Scientific instruments of the seventeenth and eighteenth centuries and their makers.* London: Batsford, 1972. Translation of *Les instruments scientifiques aux XVII et XVIII siècles.* Paris, 1953.

Delaporte, François (1982). *Nature's second kingdom: Exploration of vegetality in the eighteenth century.* Cambridge: MIT Press, 1982. Translation of *Le second règne de la nature.* Paris, 1979.

Dennis, Michael Aaron (1989). "Graphic understanding: Instruments and interpretation in Hooke's *Micrographia.*" In *Science in Context* 3 (1989), pp. 309–64.

Dewhurst, Kenneth (1958). "Locke and Sydenham on the teaching of anatomy." *Medical History* 2 (1958), pp. 1–12.

———. (1980). *Thomas Willis's Oxford lectures.* Oxford: Sandford Publications, 1980.

Diepgen, Paul (1949–55). *Geschichte der Medizin: Die historische Entwicklung der Heilkunde und des ärztlichen Lebens.* 3 vols. Berlin: de Gruyter, 1949–55.

Disney, Alfred N., et al. (1928). *Origin and development of the microscope, as illustrated by catalogues of the instruments and accessories, in the collections of the Royal Microscopical Society . . .* London: Royal Microscopical Society, 1928.

Dobell, Clifford (1923). "A protozoological bicentenary: Antony van Leeuwenhoek (1632–1723) and Louis Joblot (1645–1723)." *Parasitology* 15 (1923), pp. 308–19.

———. (1932). *Antony van Leeuwenhoek and his "little animals."* London: Bale, 1932.

Duchesneau, François (1975). "Malpighi, Descartes, and the epistemological problems of iatromechanism." In *Reason, experiment and mysticism in the scientific revolution.* Ed. M. L. Righini Bonelli and William R. Shea. New York: Science History Publications, 1975, pp. 111–30.

Dumaitre, Paule (1964). "Un anatomiste Espagnol à Paris au 17e siècle: Chrysostome Martinez et ses rarissimes planches d'anatomie." *Médecine de France* nr. 154 (1964), pp. 10–15.

Egerton, Frank N. (1968). "Leeuwenhoek as a founder of animal demography." *Journal of the History of Biology* 1 (1968), pp. 1–22.

Espinasse, Margaret (1956). *Robert Hooke.* London: Heinemann, 1956.

Farber, Eduard (1970). "Borel, Pierre." In *Dictionary of scientific biography.* Ed. Charles G. Gillispie. New York: Scribner's, 1970, vol. 2, pp. 305–6.

Farley, John (1972). *The spontaneous generation controversy from Descartes to Oparin.* Baltimore: Johns Hopkins University Press, 1977.

Fodera Serio, Giorgia, et al. (1983). "Light, colors and rainbow in Giovan Battista Hodierna (1597–1660)." *Annali dell'Istituto e Museo di Storia della Scienza di Firenze* 8 (1983), pp. 59–75.

Folter, Rolf J. de (1978). "A newly discovered Oeconomia Animalis, by Pieter Muis of Rotterdam (c. 1645–1721)." *Janus* 65 (1978), pp. 183–204.

Ford, Brian J. (1985). *Single lens: The story of the simple microscope.* London: Heinemann, 1985.

Fournier, Marian (1981). "Huygens' microscopical researches." *Janus* 68 (1981), pp. 199–210.

———. (1985). "De microscopische anatomie in Bidloo's Anatomia humani corporis (1685)." *Tijdschrift voor de geschiedenis der geneeskunde, natuurwetenschappen, wiskunde en techniek* 8 (1985), pp. 187–208.

———. (1989). "Huygens' designs for a simple microscope." *Annals of Science* 46 (1989), pp. 575–96.

Franceschini, Pietro (1970). "Buonanni, Filippo." In *Dictionary of scientific biography.* Ed. Charles Gillispie. New York: Scribner's, 1970, vol. 2, pp. 591–92.

Frank, Robert G. (1979). "The image of Harvey in Commonwealth and restoration England." In *William Harvey and his age.* Ed. Jerome J. Bylebyl. Baltimore: Johns Hopkins University Press, 1979, pp. 103–43.

———. (1980). *Harvey and the Oxford physiologists: A study of scientific ideas.* Berkeley: University of California Press, 1980.

Frison, Edward (1965). "Bij de driehonderdste verjaring van Robert Hooke's "Micrographia" (1665–1965): De opgang van de microscopie in Engeland in de tweede helft der 17de eeuw." *Scientiarum historia* 7 (1965), pp. 92–104.

Gasking, Elisabeth B. (ca. 1962). *Investigations into generation 1651–1828*. Baltimore: Johns Hopkins University Press.

Glick, Thomas F. (1974). "Martinez, Crisostomo." In *Dictionary of scientific biography*. Ed. Charles G. Gillispie. New York: Scribner's, 1974, vol. 9, pp. 145–46.

Gloede, Wolfgang (1986). *Vom Lesestein zum Elektronenmikroskop*. Berlin: VEB Verlag, 1986.

Govi, Gilberto (1889). "The compound microscope invented by Galileo." *Journal of the Royal Microscopical Society* 9 (1889), pp. 574–98.

Gray, Frieda, and Peter Gray (1956). *Annotated bibliography of works in Latin alphabet languages on biological microtechnique*. Dubuque: Brown Company, 1956.

Grmek, Mirko D. (1967). "Réflexions sur des interprétations mécanistes de la vie dans la physiologie du 17e siècle." *Episteme* 1 (1967), pp. 17–30.

———. (1970a). "Baglivi, Georgius." In *Dictionary of scientific biography*. Ed. Charles G. Gillispie. New York: Scribner's, 1970, vol. 1, pp. 391–92.

———. (1970b). "La notion de fibre vivante ches les médecins de l'école iatrophysique." *Clio medica* 5 (1970), pp. 297–318.

Grondona, Felice (1964). "Il 'De renibus' di Marcello Malpighi." *Physis* 6 (1964), pp. 385–431.

Guerrini, Anita (1985). "James Keill, George Cheyne, and Newtonian physiology, 1690–1740." *Journal of the History of Biology* 18 (1985), pp. 247–66.

———. (1987). "Archibald Pitcairne and Newtonian medicine." *Medical History* 31 (1987), pp. 70–83.

———. (1989). "The ethics of animal experimentation in seventeenth-century England." *Journal of the History of Ideas* 50 (1989), pp. 391–407.

Gunther, Robert T. (1930). *Early science in Oxford: The life and work of Robert Hooke,* vol. 7. Oxford: Printed for the author, 1930.

Haaxman, Pieter J. (1875). "Antony van Leeuwenhoek: De ontdekker der infusorien, 1675–1875." Leiden: van Doesburgh, 1875.

Halbertsma, Hiddo (1843). *Dissertatio historico-medica inauguralis de Antonii Leeuwenhoeckii meritis in quasdam partes anatomiae microscopicae . . .* Deventer: J. de Lange, 1843.

Hall, A. Rupert (1963). *From Galileo to Newton: 1630–1720*. London: Collins, 1963.

———. (1966). *Hooke's Micrographia 1665–1965*. London: University of London, 1966.

Hall, A. Rupert, and Marie Boas Hall, editors. (1965–1986). *The correspondence of Henry Oldenburg*. 13 vols. Madison: University of Wisconsin Press, 1965–1986.

Hall, Diane L. (1981). *Why do animals breathe?* New York: Arno Press, 1981. Reprint of "From Mayo to Haller." Thesis, Yale University, 1966.

Hall, Mary Boas (1966). *Experimental philosophy . . .* Sources of Science 21. New York: Johnson Reprint Co., 1966.

———. (1991). *Promoting experimental learning: experiment and the Royal Society 1660–1727.* Cambridge: Cambridge University Press, 1991.

Hall, Thomas S. (1951). *A source book in animal biology.* New York: McGraw–Hill, 1951.

———. (1969). *Ideas of life and matter: Studies in the history of general physiology 600 B.C.–1900 A.D.* 2 vols. Chicago: University of Chicago Press, 1969.

———. (1970). "Descartes' physiological method: Position, principles, examples." *Journal of the History of Biology* 3 (1970), pp. 53–79.

Harting, Pieter (1876). *Gedenkboek van het den 8e September 1875 gevierde 200-jarig herinneringsfeest der ontdekking van de mikroskopische wezens, door Antony van Leeuwenhoek.* The Hague, 1876.

Harwood, John T. (1989). "Rhetoric and graphic in Micrographia." In *Robert Hooke: New Studies.* Ed. Michael Hunter and Simon Schaffer. Woodbridge: Boydell Press, 1989, pp. 119–47.

Hayman, J. M. (1925). "Malpighi's 'Concerning the structure of the kidney's': A translation and introduction." *Annals of Medical History* 7 (1925), pp. 242–63.

Helden, Albert van (1977a). "The development of compound eyepieces, 1640–1670." *Journal of the History of Astronomy* 8 (1977), pp. 26–37.

———. (1977b). *The invention of the telescope.* Transactions of the American Philosophical Society, 67, part 4. Philadelphia: American Philosophical Society, 1977.

———. (1983). "The birth of the modern scientific instrument, 1550–1700." In *The uses of science in the age of Newton.* Ed. John G. Burke. Berkeley: University of California Press, 1983.

Heniger, Johannes (1973). "Leeuwenhoek, Antoni van." In *Dictionary of scientific biography.* Ed. Charles G. Gillispie. New York: Scribner's, 1973, vol. 8, pp. 126–30.

Henry, John (1989). "Robert Hooke, the incongruous mechanist." In *Robert Hooke: New studies.* Ed. Michael Hunter and Simon Schaffer. Woodbridge: Boydell Press, 1989, pp. 149–80.

Hesse, Mary B. (1966). "Hooke's philosophical algebra." *Isis* 57 (1966), pp. 67–83.

Hooykaas, Reyer (1950). "Antonie van Leeuwenhoek's kristalmoleculen." *Chemisch Weekblad* 46 (1950), pp. 441–42.

Hoppe, Brigitte (1976). *Biologie: Wissenschaft von der belebten Materie von der Antike zur Neuzeit: Biologische Methodologie und Lehren von der stofflichen Zusammensetzung der Organismen.* Sudhoffs Archive, Beiheft 17. Wiesbaden: Steiner, 1976.

Houghton, Bernard (1975). *Scientific periodicals, their historical development, characteristics and control.* London: Bingley, 1975.

Hughes, Arthur (1958). "Peter Mundy, the first English microscopist." *Journal of the Royal Microscopical Society,* series 3, 78 (1958), pp. 74–76.

Humbert, Pierre (1951). "Peiresc et le microscope." *Revue d'histoire des sciences et de leurs applications* 4 (1951), pp. 154–58.

Hunter, Michael (1981). *Science and society in restoration England.* Cambridge: Cambridge University Press, 1981.

———. (1982). "Early problems in professionalizing scientific research: Nehemiah Grew (1641–1712) and the Royal Society, with an unpublished letter to Henry Oldenburg." *Notes and Records of the Royal Society of London* 36 (1982), pp. 189–209.

Jahn, Ilse, et al. (1982). *Geschichte der Biologie: Theorien, Methoden, Institutionen, Kurzbiographien . . .* Jena: Fischer, 1982.

Keller, Alex G. (1974). "Perrault, Pierre." In *Dictionary of scientific biography.* Ed. Charles G. Gillispie. New York: Scribner's, 1974, vol. 10, pp. 519–21.

Keynes, Geoffrey (1960). *A bibliography of Dr. Robert Hooke.* Oxford: Clarendon Press, 1960.

Kidwell, Clara S. (1970). "The Accademia dei Lincei and the Apiarum: A case study of the activities of a seventeenth century scientific society." Thesis, University of Oklahoma, 1970.

King, Lester S. (1965). *The background of Herman Boerhaave's doctrines.* Boerhaave Lecture, 17 September 1964. Leiden: Universitaire Pers, 1965.

Kisch, Arnold I. (1951). "Predecessors of the glass micrometer." *Journal of the Royal Microscopical Society* 71 (1951), pp. 181–85.

Kronick, David A. (1962). *A history of scientific and technical periodicals: The origin and development of the scientific and technological press 1665–1790.* New York: Scarecrow Press, 1962.

Lechevalier, Hubert (1976). "Louis Joblot and his microscopes." *Bacteriological Reviews* 40 (1976), pp. 241–58.

Leersum, Evert C. van (1927). "Inleiding." In *Opuscula selecta Neerlandicorum de arte medica.* Amsterdam: Sumptibus Societatis, 1927, vol. 6, VII–XV.

Lefanu, William (1990). *Nehemiah Grew M.D., F.R.S.: A study and bibliography of his writings.* Winchester: St. Paul's Bibliographies, 1990.

Lindeboom, Gerrit A. (1973). *Reinier de Graaf, leven en werken: 30-7-1641/17-8-1673.* Delft: Elmar, 1973.

———. (1975a). "Het Collegium Privatum Amstelodamense (1664–1673)." *Nederlands Tijdschrift voor Geneeskunde* 119 (1975), pp. 1248–54.

———. (1975b). "Dog and frog: Physiological experiments at Leiden during the seventeenth century." In *Leiden University in the seventeenth century: An exchange of learning.* Ed. Th. H. Lunsingh Scheurleer et al. Leiden: Brill, 1975, pp. 279–93.

———. (1975c). *The letters of Jan Swammerdam to Melchisedec Thévenot.* Amsterdam: Swets and Zeitlinger, 1975.

———. (1980). "Zeer kleine glasbolletjes als sterk vergrotende mikroscoopjes gebruikt door Nederlanders in de tweede helft der zeventiende eeuw." In *Zusammenhang: Festschrift fur Marielene Putscher.* Ed. Otto Baur and Otto Glandien. Cologne: Wienand, 1984, pp. 337–51.

———. (1981). "Jan Swammerdam als microscopist." *Tijdschrift voor de geschiedenis der geneeskunde, natuurwetenschappen, wiskunde en techniek* 4 (1981), pp. 87–110.

———. (1982a). "Jan Swammerdam (1637–1680) and his Biblia Naturae." *Clio medica* 17 (1982), pp. 113–31.

———. (1982b). "Leeuwenhoek and the problem of sexual reproduction." In *Antoni van Leeuwenhoek 1632–1723: Studies on the life and work.* Ed. L. C. Palm and H. A. M. Snelders. Amsterdam: Rodopi, 1982, pp. 129–52.

———. (1983). "Anton de Heide als proefondervindelijk onderzoeker." *Tijdschrift voor de geschiedenis der geneeskunde, natuurwetenschappen, wiskunde en techniek* 6 (1983), pp. 122–34.

Lopez Piñero, José M. (1982). *El atlas anatomico de Crisostomo Martinez: Grabador y microscopista del siglo 17: Estudio y transcripcion de José Maria Lopez Piñero.* Ayuntamiento de Valencia: Artes Graficas Soler, 1982.

Luyendijk-Elshout, Antonie M. (1974). "Antony Nuck (1650–1692): The 'Mercator' of the body fluids." In *Circa Tiliam: Studia historiae medicinae Gerrit Arie Lindeboom septuagenario oblata.* Leiden: Brill, 1974, pp. 150–64.

———. (1975). "Oeconomia animalis, pores and particles: The rise and fall of the Mechanical Philosophical School of Theodoor Craanen." In *Leiden University in the seventeenth century: An exchange of learning.* Ed. Th. H. Lunsingh Scheurleer et al. Leiden: Universitaire Pers, 1975, pp. 295–307.

Man, Johannes C. de (1905). *Antonius de Heide Med. Doctor te Middelburg, ontdekker der later zoo beroemd geworden trilhaarbeweging.* Middelburg: Kröber, 1905.

Martin, Hubert de (ca. 1970). *Griendel von Ach: Ein Mikroskopiker der Barockzeit.* Vienna: Höhere Graphische Bundes-Lehr- und Versuchsanstalt.

———. (1973). *Vademecum mikroskopischer Literatur: Ein Verzeichnis selbständig erschienener Werke über Mikroskopie von 1590–1970.* Vienna: Höhere Graphische Bundes-Lehr- und Versuchsanstalt, 1973.

Martin, Hubert de, and Waltraud de Martin (1983). *Vier Jahrhunderte Mikroskop.* Wiener Neustadt: Weiburg, 1983.

Matsuo, Yukitoshi (1975). "Plant physiology, especially the motion of the sap, in the early days of the Royal Society of London." *Proceedings no. 3 [of the] 14th International Congress of the History of Science, Tokyo and Kyoto, 19–27 August 1974.* Tokyo: Science Council of Japan, 1975, pp. 47–50.

Meinel, Christopher (1988). ""Das letzte Blatt im Buch der Natur": Die Wirklichkeit der Atome und die Antinomie der Anschauung in den Korpuskulartheorien der frühen Neuzeit." *Studia Leibnitiana* 20 (1988), pp. 1–18.

Metcalfe, Charles R. (1972). "Grew, Nehemiah." In *Dictionary of scientific biography.* Ed. Charles Gillispie. New York: Scribner's, 1972, vol. 5, pp. 534–36.

Meyer, A. W. (1938). "Leeuwenhoek as experimental biologist." *Osiris* 3 (1938), pp. 103–22.

Miller, Genevieve (1968). "Leeuwenhoek's observations on the blood and capillary vessels." In *Medicine, science and culture.* Ed. Lloyd G. Stevenson and Robert P. Multhauf. Baltimore: Johns Hopkins University Press, 1968, pp. 114–22.

———. (1981). "Early concepts of the microvascular system: William Harvey to Marshall Hall, 1628–1831." In *The analytic spirit.* Ed. Harry Woolf. Ithaca: Cornell University Press, 1981, pp. 257–78.

Möbius, Martin (1901). *Marcello Malpighi: Die Anatomie des Pflanzen.* Leipzig: Engelmann, 1901.

Montagu, M. F. Ashley (1943). *Edward Tyson, M.D., F.R.S. 1650–1708 and the rise of human and comparative anatomy in England: A study in the history of science.* Philadelphia: American Philosophical Society, 1943.

Moravia, Sergio (1978). "From homme machine to homme sensible: Changing eighteenth-century models of man's image." *Journal of the History of Ideas* 39 (1978), pp. 45–60.

Nachet, Albert (1929). *Collection Nachet: Instruments scientifiques et livres anciens. Notice sur l'invention du microscope et son évolution. Liste de savants, constructeurs et amateurs du 16e au milieu du 19e siècle.* Paris: Georges Petit, 1929.

Needham, Joseph (1934). *A history of embryology.* 2d ed. Revised with the assistance of Arthur Hughes. Cambridge: Cambridge University Press, 1959.

Nelson, Edward M. (1902). "A bibliography of works (dated not later than 1700) dealing with the microscope and other optical subjects." *Journal of the Royal Microscopical Society* 22 (1902), pp. 20–23.

Nicolson, Marjorie H. (1956). *Science and imagination.* Ithaca: Great Seal, 1956.

Nordenskiöeld, Erik (1928). *The history of biology.* New York: Knopf, 1928. Translation of *Biologins Historia.* Stockholm, 1920–24.

Nordströem, Johan (1955). "Swammerdamiana: Excerpts from the travel journal of Olaus Borrichius and two letters from Swammerdam to Thévenot. Together with an appendix: the history of Swammerdam's demonstration of the valves in the lymphatic vessels." *Lychnos* (1954–55), pp. 21–65.

Nowak, Hans Peter (1984). *Geschichte des Mikroskops: Eine Ausstellung der medizinhistorischen Sammlung der Universität Zürich.* Zurich: Medizinhistorisches Institut der Universität Zürich, 1984.

O'Brien, Gordon W. (1975). "Power, Henry." In *Dictionary of scientific biography.* Ed. Charles G. Gillispie. New York: Scribner's, 1975, vol. 11, pp. 121–22.

Oldroyd, David R. (1972). "Robert Hooke's methodology of sciences as exemplified in his 'Discourse of earthquakes.'" *British Journal for the History of Science* 6 (1972), pp. 109–30.

———. (1987). "Some writings of Robert Hooke on procedures for the prosecution of scientific inquiry, including his 'Lectures of things requisite to a natural history.'" *Notes and Records of the Royal Society of London* 41 (1987), pp. 145–67.

Ornstein, Martha (1938). *The role of scientific societies in the seventeenth century.* Chicago: University of Chicago Press, 1938.

Palm, Lodewijk C. (1978). "Antoni van Leeuwenhoek en de ontdekking der haarvaten." *Tijdschrift voor de geschiedenis der geneeskunde, natuurwetenschappen, wiskunde en techniek* 1 (1978), pp. 170–77.

———. (1982). "Antoni van Leeuwenhoek: Malacological researches as an example of his biological studies." In *Antoni van Leeuwenhoek 1632–1723 . . .* Ed. L. C. Palm and H. A. M. Snelders. Amsterdam: Rodopi, 1982, pp. 153–67.

———. (1989). "Italian influences on Antoni van Leeuwenhoek." In *Italian scientists in the low countries in the 17th and 18th centuries.* Ed. C. S. Maffioli and L. C. Palm. Amsterdam: Rodopi, 1989, pp. 147–63

Palm, Lodewijk C., and Henricus A. M. Snelders (1982). *Antoni van Leeuwenhoek 1632–1723: Studies on the life and work of the Delft scientist commemorating the 350th anniversary of his birthday.* Amsterdam: Rodopi, 1982.

Pas, Peter W. van der (1970). "A note on the origin of staining techniques for microscopical preparations." *Scientiarum historia* 12 (1970), pp. 63–72.

Peyer, Bernhard (1946). *Die biologische Arbeiten des Arztes Johannes von Muralt 1645–1733.* Thayngen: Stiftung von Schnyder von Wartensee, 1946.

Pighetti, Clelia (1961). "Giovan Battista Odierna e il suo discorso su l'occhio della Mosca." *Physis* 3 (1961), pp. 309–35.

Ploeg, Willem (1934). *Constantijn Huygens en de natuurwetenschappen.* Rotterdam: Nijgh en Van Ditmar, 1934. Reprint, thesis, Leiden, 1934.

Pyle, Andrew J. (1987). "Animal generation and the mechanical philosophy: Some light on the role of biology in the scientific revolution." *History and Philosophy of the Life Sciences* 9 (1987), pp. 225–54.

Ritterbush, Philip C. (1964). *Overtures to biology: The speculations of eighteenth-century naturalists.* New Haven: Yale University Press, 1964.

Roger, Jacques (1971). *Les sciences de la vie dans la pensée Française du 18e siècle: La génération des animaux de Descartes à l'Encyclopédie.* 2d ed. Paris: Armand Collin, 1971.

Rooseboom, Maria (1939). "Concerning the optical qualities of some microscopes made by Leeuwenhoek." *Journal of the Royal Microscopical Society* 59 (1939), pp. 177–83.

———. (1950). "Leeuwenhoek, the Man: A son of his nation and his time." *Bulletin of the British Society for the History of Science* 1 (1950), pp. 79–85.

———. (1953). "Influences of the invention of the microscope on biological thought." *Actes du 7e Congrès international d'histoire des sciences Jérusalem 4–12 Août 1953*. Paris: Académie Internationale d'Histoire des Sciences, 1953, pp. 522–30.

———. (1956a). "The introduction of mounting media in microscopy and their importance for biological science." *Actes du 8e Congrès Internationale d'histoire des sciences Florence 3–9 September 1956*. Florence: Gruppo Italiano di Storia delle Scienze, 1958, pp. 602–7.

———. (1956b). *Microscopium*. Mededeling no. 95. Leiden: Rijksmuseum voor de Geschiedenis der Natuurwetenschappen, 1956.

———. (1958). "Christiaan Huygens et la microscopie." *Archives Néerlandaises de Zoologie* 13 (1958), 1st supplement, pp. 59–73.

———. (1959). "Antoni van Leeuwenhoek vu dans le milieu scientifique de son époque." *Archives internationales d'histoire des sciences* 12 (1959), pp. 27–46.

———. (1967). "The history of the microscope." *Proceedings of the Royal Microscopical Society* 2 (1967), pp. 266–93.

Rothschuh, Karl E. (1953). *Geschichte der Physiologie*. Berlin: Springer, 1953.

Ruestow, Edward G. (1980). "The rise of the doctrine of vascular secretion in The Netherlands." *Journal of the History of Medicine* 35 (1980), pp. 265–87.

———. (1983). "Images and ideas Leeuwenhoek's perception of the spermatozoa." *Journal for the History of Biology* 16 (1983), pp. 185–224.

———. (1984). "Leeuwenhoek and the campaign against spontaneous generation." *Journal of the History of Biology* 17 (1984), pp. 225–48.

———. (1985). "Piety and the defense of natural order: Swammerdam on generation." In *Religion, science and worldview: Essays in honor of Richard S. Westfall*. Ed. Margaret J. Osler and Paul Lawrence Farber. Cambridge: Cambridge University Press, 1985, pp. 217–41.

Scherz, Gustav (1968). *Steno and brain research in the seventeenth century. Proceedings of the International historical symposium on Nicolaus Steno and brain research in the seventeenth century, Copenhagen 18–20 August 1965*. Ed. Gustav Scherz. Oxford: Pergamon Press, 1968.

———. (1976). "Stensen, Niels." In *Dictionary of scientific biography*. Ed. Charles G. Gillespie. New York: Scribner's, 1976, vol. 13, pp. 30–35.

Schierbeek, Abraham (1939). "Leeuwenhoeck en zijn globulentheorie." *Natuurwetenschappelijk Tijdschrift* 21 (1939), pp. 185–89.

———. (1947). *Jan Swammerdam (12 Februari 1637–17 Februari 1680): Zijn leven en zijn werken. Met een hoofdstuk De genealogie van Swammerdam en verdere archivalia door H. Engel*. Lochem: De Tijdstroom, 1947.

———. (1950–51). *Antoni van Leeuwenhoek, zijn leven en zijn werken*. 2 vols. Lochem: De Tijdstroom, 1950–51.

Schulte, Bento P. M. (1968). "Swammerdam and Steno." In *Steno and brain research.* Ed. Gustav Scherz. Oxford: Pergamon Press, 1968, pp. 35–41.

Seters, Wouter H. van (1933). "Leeuwenhoecks microscopen, praepareer- en observatiemethodes." *Bijdragen tot de geschiedenis der geneeskunde* 13 (1933), pp. 217–35.

———. (1982). "Can Antoni van Leeuwenhoek have attended school at Warmond?" In *Antoni van Leeuwenhoek 1632–1723: Studies on the life and work.* Ed. L. C. Palm and H. A. M. Snelders. Amsterdam: Rodopi, 1982, pp. 3–11.

Settle, Thomas B. (1970). "Borelli, Giovanni Alfonso." In *Dictionary of scientific biography.* Ed. Charles G. Gillispie. New York: Scribner's, 1970, vol. 2, pp. 306–14.

Shapin, Steven (1989). "Who was Robert Hooke?" In *Robert Hooke: New Studies.* Ed. Michael Hunter and Simon Schaffer. Woodbridge: Boydell Press, 1989, pp. 253–85.

Singer, Charles (1914). "Notes on the early history of microscopy." *Proceedings of the Royal Society of Medicine* 7 (1914), pp. 247–79.

———. (1915). "The dawn of microscopical discovery." *Journal of the Royal Microscopical Society* (1915), pp. 317–40.

———. (1931). *A short history of biology: A general introduction to the study of living things.* Oxford: Clarendon Press, 1931.

———. (1953). "The earliest figures of microscopical objects." *Endeavour* 12 (1953), pp. 197–201.

———. (1955). "The first English microscopist Robert Hooke (1635–1703)." *Endeavour* 14 (1955), pp. 12–18.

Singer, Dorothy W. (1956). "A pioneer microscopic dissection l'Occhio della Mosca." *Actes du 8e Congrès international d'histoire des sciences, Florence-Milan 3–9 Septembre 1956.* Florence: Gruppo Italiano di Storia delle Scienze, 1958, pp. 730–34.

Smit, Pieter (1981). "Jan Swammerdam (1637–1680) und seine Beobachtungen zur Metamorphose der Insekten." In *Hallesche Physiologie im Werden: Hallesches Symposium 1981.* Ed. Wolfram Kaiser and Hans Hübner. Halle: Martin-Luther-Universität, 1981, pp. 35–43.

Smit, Pieter, and Johannes Heniger (1975). "Antoni van Leeuwenhoek (1632–1723) and the discovery of the bacteria." *Antonie van Leeuwenhoek* 41 (1975), pp. 217–28.

Snelders, Henricus A. M. (1982). "Antoni van Leeuwenhoek's mechanistic view of the world." In *Antoni van Leeuwenhoek 1632–1723.* Ed. L. C. Palm and H. A. M. Snelders. Amsterdam: Rodopi, 1982, pp. 57–78.

Snow Miller, William (1922). "Thomas Willis and his 'De Phthisi Pulmonari.'" *American Review of Tuberculosis* 5 (1922), pp. 934–49.

Star, Pieter van der (1953). *Descriptive catalogue of the simple microscopes in the Rijksmuseum voor de Geschiedenis der Natuurwetenschappen (National*

Museum of the History of Science) at Leyden. Communication no. 87. Leiden: Rijksmuseum voor de Geschiedenis der Natuurwetenschappen, 1953.

Stenn, Frederik (1941). "Giorgio Baglivi." *Annals of Medical History*, 3d series, 3 (1941), pp. 183–94.

Stroup, Alice (1990). *A company of scientists: Botany, patronage, and community at the seventeenth-century Parisian Royal Academy of Sciences*. Berkeley: University of California Press, 1990.

Tierie, Gerrit (1932). *Cornelis Drebbel (1572–1633)*. Amsterdam: H. J. Paris, 1932.

Turner, Gerard l'E. (1969). "The history of optical instruments: A brief survey of sources and modern studies." *History of Science* 8 (1969), pp. 53–93. Reprinted in G. l'E. Turner, *Essays on the history of the microscope*. Oxford: Senecio, 1980.

———. (1972). "Micrographia historica: The study of the history of the microscope." *Proceedings of the Royal Microscopical Society* 7 (1972), pp. 120–49. Reprinted in G. l'E. Turner, *Essays on the history of the microscope*. Oxford: Senecio, 1980.

———. (1974). "Microscopical communication." *Journal of Microscopy* 100 (1974), pp. 3–20.

———. (1980). *Essays on the history of the microscope*. Oxford: Senecio, 1980.

———. (1981). *Collecting microscopes*. New York: Mayflower Books, 1981.

———. (1985). "Animadversions on the origins of the microscope." In *The light of nature: Essays in the history and philosophy of science presented to A. C. Crombie*. Ed. John D. North and J. J. Roche. Dordrecht: Nijhoff, 1985, pp. 193–207.

Visser, Robert P. W. (1981). "Theorie en praktijk van Swammerdams wetenschappelijke methode in zijn entomologie." *Tijdschrift voor de geschiedenis der geneeskunde, natuurwetenschappen, wiskunde en techniek* 4 (1981), pp. 63–73.

Webster, Charles (1967). "Henry Power's experimental philosophy." *Ambix* 14 (1967), pp. 150–78.

Westfall, Richard S. (1972). "Hooke, Robert." In *Dictionary of scientific biography*. Ed. Charles G. Gillispie. New York: Scribner's, 1972, vol. 6, pp. 481–88.

Williams, Wesley C. (1976). "Tyson, Edward." In *Dictionary of scientific biography*. Ed. Charles G. Gillispie. New York: Scribner's, 1976, vol. 13, pp. 526–28.

Wilson, Catherine (1988). "Visual surface and visual symbol the microscope and the occult in early modern science." *Journal of the History of Ideas* 49 (1988), pp. 85–108.

Wilson, Leonard G. (1960). "The transformation of ancient concepts of respiration in the seventeenth century." *Isis* 51 (1960), pp. 161–72.

———. (1961). "William Croone's theory of muscle contraction." *Notes and Records of the Royal Society of London* 16 (1961), pp. 158–78.

Winsor, Mary P. (1976). "Swammerdam, Jan." In *Dictionary of scientific biography.* Ed. Charles G. Gillispie. New York: Scribner's, 1976, vol. 13, pp. 168–75.

Wolfe, David E. (1961). "Sydenham and Locke on the limits of anatomy." *Bulletin of the History of Medicine* 35 (1961), pp. 193–220.

Woodruff, Lorande L. (1937). "Louis Joblot and the protozoa." *Scientific monthly* 44 (1937), pp. 41–47.

Young, James (1929). "Malpighi's 'De Pulmonibus.'" *Proceedings of the Royal Society of Medicine, Section of the History of Medicine* (1929), pp. 1–11.

Zanobio, Bruno (1960). "L'immagine filamentoso–reticolare nell'anatomia microscopica dal 17 al 19 secolo." *Physis* 2 (1960), pp. 299–317.

———. (1971). "Micrographie illusoire et théories sur la structure de la matière vivante." *Clio medica* 6 (1971), pp. 25–40.

Zirkle, Conway (1965). *Nehemiah Grew: The anatomy of plants.* Reprinted from the 1682 edition. New York: Johnson Reprint Co., 1965.

Zuylen, Jan van (1981a). "The microscopes of Antoni van Leeuwenhoek." *Journal of Microscopy* 121 (1981), pp. 309–28. Reprinted in L. C. Palm and H. A. M. Snelders, eds., *Antoni van Leeuwenhoek 1632–1723*. Amsterdam: Rodopi, 1982, pp. 29–55.

———. (1981b). "On the microscopes of Antoni van Leeuwenhoek." *Janus* 68 (1981), pp. 159–98.

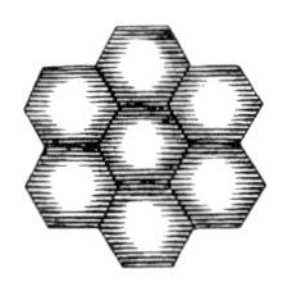

Index

Académie des Sciences, 22, 41, 136, 168
Accademia dei Lincei, 11, 25
Accademia del Cimento, 56
Accademia Fisicomatematica Romana, 19
alimentary duct, 39, 137–38, 141
amateur of science, 5
analogy: between animals and plants, 60, 77, 105, 120; between insects and higher animals, 151; with mechanical models, 55, 85, 102, 108, 124, 161; reasoning from, 59, 85, 120
anatomia subtilis, 32, 55, 58, 187
anatomical atlas, 138–41
anatomical method, 104, 187
anatomy, comparative, 57–58, 76, 141; of animals and man, 135–46; microscopic, 138–40. *See also* insects: anatomy of
animal spirits, 95, 106–7
Anonymous (researchers), 172, 176
aphids, 90
Asylus, 70–71
atomical hypothesis. *See* corpuscular philosophy

B

Baconianism. *See* empirical method
bacteria, 169
Baglivi, G., 111, 129
Baker, H., 43
Balthasar, T., 35
Bartholin, T., 136
beard of oats, 54, 102
bee, 15, 148; compound eye, 148–50, 162
Beeckman, I., 26
beetle, 164
Bellini, L., 190
bibliometry, 40–44
Bidloo, G., 37–39, 138–40, 156
Blaes, G., 65, 141
Blankaart, S., 141
blood, 33, 88–89, 129, 146
Boerhaave, H., 64, 68, 80, 131
Bombyx, 115–16
bones, 138
Bonnet, C., 4
Bonomo, G. C., 156
Bontekoe, C., 133
Borch, O., 62, 156
Borel, P., 14, 27–29, 99, 191
Borelli, G. B., 56, 61, 109, 129
Bourignon, A. de, 64
Boyle, R., 74, 78, 93–94, 130
brain, 54, 113–14, 140
Browne, T., 26
Buonanni, F., 19, 29, 154, 191
Butterfield, M., 22

C

Campani, G., 18, 19
capillaries, 89, 112, 129–34
Catelan, 156
Celio, M., 19
Cesi, F., 11, 25, 191
Cestoni, G., 156
Chain of Being, 169, 177
Chapotôt, L., 22
Charleton, W., 27, 93, 94, 186
chemistry, of plants, 74, 78–79
Chérubin d'Orleans, 36, 37
chicken embryo, 95, 116
circulation: of blood, 129, 131; of plant sap, 124–26
classification, of living organisms, 195
Cock, C., 13
Collegium Naturae Curiosorum, 41
Collegium privatum Amstelodamense, 65, 141, 146
coloring agents, 32
communication, of microscopical data, 37–39
comparative anatomy. *See* anatomy, comparative
compound eye, 147, 156–57, 159; bee, 148–50, 162; dragonfly, 157, 162–64; fly, 16, 26–27, 99–100; vision mechanism, 99–101, 147, 156–57
constancy of scale in nature, 166
cork, 30–31
corpuscular philosophy, 77, 179–81, 188; and microscopy, 95–96, 186; theory of living matter, 87, 128
Craanen, T., 132
Croone, W., 106–9, 129
Cuff, J., 18
Culpeper, E., 18–19, 24, 159
Cuno, C. C., 22

D

Depouilly, 22
Descartes, R., and physiology, 87–88, 104–6, 110, 114, 188
Divini, E., 12, 14
dragonfly, 157, 162–64
Drebbel, C., 10
drone fly, 99–100
Duverney, J. G., 55

E

Ellis, J., 4
empirical method, 50, 65–66, 75, 155, 187–88
English Physiologists, 130
epigenesis, 71
esophagus. *See* alimentary duct
eye, compound. *See* compound eye
eye, vertebrate, 161–63

F

fibers, 88, 128–34, 193
fly. *See* compound eye: fly
Fontana, F., 24–26

G

Galileï, G., 1, 9–10, 26
Garden, G., 85, 86
Gassendi, P., 92, 94
generation, 69, 82–83, 85–86, 90, 153; of microorganisms, 173–77; spontaneous, 68, 90. *See also* mechanical philosophy: and generation
glandular bodies, 112–14
Glanvill, J., 27
Glisson, F., 55
Graaf, R. de, 65, 81–82, 105, 136, 182
Great Chain of Being, 169, 177
Gresham College, 50

Grew, N., 72–79, 121–28; analogical reasoning, 77, 122–24; animals, 76–77, 122; and Boyle, 74; chemistry, 74–75, 77–79, 127, 179; and Hooke, 72, 74, 182; and Leeuwenhoek, 86, 182–83; and Malpighi, 73, 118, 182; microorganisms, 97, 167; microtechnique, 30, 122; organic matter, 74, 77–79, 127–28; physicotheology, 75, 195; plant structure, 86, 121–24, 127, 129, 179, 183; and Royal Society, 72–73, 128; sap transport, 124–26; scientific research, 72, 75–76, 128, 179; vegetation, 77, 124–27, 180
Griendel von Ach, J. F., 13–14, 153, 191
growth, 53, 69, 71, 117–18
gut. *See* alimentary duct

H

Hartsoeker, N., 22, 24, 81, 168
Harvey, W., 129
Hedwig, J., 193
Heide, A. de, 130–31, 143, 145
Hertel, C. G., 19
Hevelius, J., 18
Hire, P. de la, 156, 168
Hoffmann, F., 131
Hooke, R., 49–55, 96–103; compound eye, 99–101, 150, 156; corpuscular hypothesis, 179–80, 188; generation, 54–55; and Grew, 72, 74, 182; hygroscope, 102; and Leeuwenhoek, 97, 110; and Malpighi, 182; mechanical view of nature, 52–55, 102–3, 108, 179; *Micrographia*, 49, 96–98; micrometry, 34; microorganisms, 97, 167–68; microscopes, 12–13, 18, 97, 159, 189; microscopes, value of, 1, 97, 101–2, 169, 195–96; microscopic illustrations, 38; microscopic images, 35–36, 98; microscopy, decline of, 2; microtechnique, 30, 33–34, 97; muscle, 107–8, 110; organic matter, 53–54, 180; scientific enquiry, 50–52, 98; scientific instruments, 50–52, 178–79; "Philosophical Algebra," 52–54; plants, 52–54, 74, 86, 102, 183; and Royal Society, 49–50, 73, 96
Horne, J. van, 65
Hudde, J., 13, 147
Huygens, C. (Jr.), 19, 22, 168
Huygens, C. (Sr.), 80, 168
Huygens, Chr., 43–44, 80–81, 191; generation of microorganisms, 174–76; microorganisms, 168–76; microscopes, 13–14, 19, 22, 157, 159; microtechnique, 34
hygroscope, 102

I

ichneumon fly, 158
infusoria. *See* microorganisms
injection, 32, 58, 133
insects: anatomy of, 94, 148, 150, 152–57, 191–92; description, 24–25, 152, 191; metamorphosis of, 68–69; trachea, function of, 86, 115. *See also* compound eye
instruments. *See* scientific instruments

J

Jansen, Z., 1
Joblot, L., 168–70, 172, 175–76
Journal des sçavans, 41, 44–45, 156
Jurin, J., 35

K

Kepler, J., 10
Kerckring, T., 36, 155
kidney, 60, 112–13
King, E., 81, 128, 136–37, 173
Kircher, A., 26, 35, 154
Kufler, J., 10, 25

L

Lairesse, G. de, 38, 138
leather-jackets, 160
Leeuwenhoek, A. van, 79–91, 157–67; analogical reasoning, 85, 89–90; bacteria, 169; capillary vessels, 129, 131; Cartesian notions, 87–88, 165, 181; compound eye, 162–64; concentric method, 82–83; generation, 82–83, 86, 90, 160; globule theory, 87, 161–62; and Grew, 86, 182–83; and Hooke, 86; and Malpighi, 86, 182; mechanical explanations, 85, 87, 161, 165–66; micrometry, 34–35, 84; microorganisms, 91, 160, 167, 169–74, 176; microscopes, 16–17, 19, 157–59; microtechnique, 32–33, 159; muscle, 107, 109–10; organic matter, 84, 87–88, 157, 161–63; personality, 79, 85–86; plants, 83, 86, 183; preformation, 90; presentation of work, 80–81, 170; quantitative approach, 84–85, 166; publications, 44–45; reception of work, 80–81, 164, 165; research style, 80, 160, 163, 165, 190; and Royal Society, 81; spermatozoa, 82–83, 90; and Swammerdam, 86, 162, 182; terminology problem, 84; uniformity in nature, 88, 90, 166; vertebrate eye, 161–62; vision mechanism, 163
Leibniz, G. W., 81, 165
Lipperhey, H., 1, 10
Lister, M., 81
liver, 112
liver fluke, 139
Lower, R., 130
lung, 60, 112, 131, 136–37
Lyonet, P., 4, 193

M

Malebranche, N., 169
Malpighi, M., 55–62, 112–21; analogical reasoning, 59–60, 117–18, 120; *anatomia subtilis*, 55, 58; animal anatomy, 55, 57, 112–13, 130–31, 136; and Borelli, 56; capillary vessels, 112, 129; chicken embryo, 116–17; generation, 116–18; glandular bodies, 112–14; and Grew, 73, 118; growth, 117–18; insects, 86, 115–16, 184, 189; and Leeuwenhoek, 182; mechanical philosophy, 61, 121, 181; mechanical physiology, 56, 60–61, 114, 121, 130–33, 189; microscopes, 29, 36, 59, 121; microscopes, limits of, 39, 59, 61, 120; microtechnique, 32–34, 59; organic matter, 60, 120, 179, 180; plants, 117–20; research, 56–58; and Royal Society, 56–57, 73, 115, 118; and Ruysch, 133; sap transport, 119–20; and Sbaraglia, 57; and Swammerdam, 184; uniformity in nature, 59, 60
Marshall, J., 18
Marsigli, L. F., 4
Martinez, C., 138
Mary, Queen of England, 80
mathematical analysis, in life sciences, 193
matter, organic. *See* organic matter, fabric of,

matter, organization of, 53–54
mechanical explanation, in life sciences, 29, 54–56, 68, 105, 165–66, 194
mechanical models, 85, 102, 108, 161
mechanical philosophy: and generation, 54–55, 69, 92; and microscopy, 4, 6, 94, 135, 186; and physiology, 29, 56, 60, 67, 106–12, 151, 181, 188, 194; and plants, 77, 105
Mentzel, C., 156
metamorphosis, 68–69
Metius, J., 1, 10
Micheli, P. A., 4
micrometry, 34–35
microorganisms, 167–77, 192; discovery, 160, 167; generation, 173–77
microscope: in anatomy, 135, 146, 187; compound, 10–13, 17–20; and corpuscular philosophy, 93–96, 186; invention of, 9–11; limits of, 11, 59, 61, 147; magnification, 13–14, 16–17; and mechanical philosophy, 4, 6, 94, 135, 186; and microscopy, rise of, 46–48, 185–86; in natural history, 191, 194–95; optical aberrations, 11–12; optics of, 11–14; in physiology, 4, 188, 190, 194; simple, 13, 19, 21–24, 47, 157, 159
microscopic anatomy, 138–40
microscopic images: fallacies of, 35–37; interpretation of, 35–36, 40, 98, 147
microscopic objects, illustration of, 25, 37–38, 138–40, 154
microscopy: decline in, 2, 5, 45, 192–95; descriptive terms in, 37, 183; heyday of, 2, 5, 45, 47, 185–92
microtechnique, 30–34, 39, 59, 147, 159
Middelburg, 1
Miscellanea curiosa, 41, 45
mold, 52–53
mole cricket, 152
Molyneux, T., 81
mosquito, 154
moth, 16
Müller, O. F., 193
Mundy, P., 27
Muralt, J. von, 45, 152, 191
muscle, 67, 106–12
mushrooms, 53
Musschenbroek, 23, 159, 183
mussel, 143, 145, 160
Muys, W. W., 32, 35, 111

N

natural history, 191, 194–95
natural theology. *See* physico-theology
nerves, 114
Newton, I., 12
nomenclature, of microorganisms, 170

O

Odierna, G. B., 14, 99; on compound eye, 24, 26–27, 29, 101, 148; microtechnique of, 27
Oldenburg, H., 56, 114, 183
order in nature, 68–69, 71
organic matter, fabric of, 54, 60, 77–79, 87–88, 120, 127, 179
Oxford Physiologists, 130

P

Panarolo, D., 27
parthenogenesis, 160
Peiresc, N-C. F. de, 10, 25–26, 191
Perrault, C., 110, 141, 156
Peter I, Tsar of Russia, 80
Philosophe, 174–76
Philosophical Transactions, 41, 44–45, 81

philosophy, corpuscular. *See* corpuscular philosophy
philosophy, mechanical. *See* mechanical philosophy
physico-theology, 28–29, 63, 75, 85, 195
physiology: chemical analysis, 105; experiments in, 58, 67; mechanical, 29, 67, 92, 104–5, 132–34, 181, 189; and microscope, 4, 188, 190, 194
Picard, J., 168
Pitcairne, A., 131–32, 134, 193
plants, 118–28; chemistry, 74, 78–79; fabric of, 119, 122–24, 128; microscopic anatomy, 31, 118–19, 122–24; transport of fluids in, 74, 119–20, 124–26, 183, 194
plenitude, principle of, 177
Poupart, F., 156, 157
Power, H., 29, 81, 92–96, 99, 187
preexistence, 69, 118
preformation, 69, 83, 90, 116–17
Puget, L., 156, 157

R

Reaumur, R-A. F. de, 4, 195
Redi, F., 142, 174
religion, and scientific research, 63–64
reproduction. *See* generation
respiration, 66–67, 130
Rømer, O., 168, 172
roundworm, 143–44
Royal Society, 3, 41, 56, 136, 181–83, 186; and Grew, 72–73, 97, 128, 167; and Hooke, 49–50, 73, 96, 167; and Leeuwenhoek, 81, 96–97, 159; and Malpighi, 56–57, 115, 118; microorganisms, 97, 167–68; muscles, 106, 108–9; and Power, 94, 187; and Swammerdam, 65
Ruysch, F., 65, 129, 132–33

S

Sachs, P. J., 29, 155, 191
Sallo, D. de, 41
Sbaraglia, G., 57
scabies, 156
Schenk, J. T., 61
Schrader, F., 155, 191
scientific instruments, 9, 50–52, 179
scientific journals, 40–41
scientific revolution, 9, 104
secretion, 131–32
sensation, 54, 113–14
silk moth, 115–16
silkworm, 184
skin, 113
snail, 95
spermatozoa, 82–83, 90
spiritus animalis, 95, 106–7
spleen, 60, 112, 139
spontaneous generation, 68, 90
Stelluti, F., 11, 14, 25–26, 191
Stensen, N., 55, 64–65, 92, 105, 107, 136
stinging nettle, 102
stomach. *See* alimentary duct
Swammerdam, J., 62–72, 146–51; anatomy of insects, 64, 68, 86, 148, 184, 191; and Blaes, 65; and Boerhaave, 64, 68; compound eye, 148, 150, 159; conception of nature, 68–69, 71; development, 68–69, 71; generation, 68–69, 71; and Hooke, 150; and Hudde, 147; on knowledge, 66–67; and Leeuwenhoek, 162, 182; at Leiden University, 62, 66–67; and Malpighi, 64, 150, 184; manuscripts, 43–44, 64; microscope, 68, 146–47; microtechnique, 147; muscle, 67, 106–7; physico-theology, 63, 195; physiology, 66–68, 151, 181; religious feelings, 63–64,

151; research, 65–66, 71, 150–51, 180; and Royal Society, 65; and Ruysch, 65; and Thévenot, 62, 64–65, 147. *See also* Collegium privatum Amstelodamense
Sydenham, T., 57
Sylvius, F. dela Boë, 67

T
tadpole, 129
tapeworm, 143
telescope, 9–10
terminology, in microscopy, 183. *See also* nomenclature, of microorganisms
Thévenot, M., 43, 62, 64–65, 147
tongue, 113
Trembley, A., 4, 45
Turquet de Mayerne, T., 26
Tyson, E., 81, 142–44, 191

U
uniformity in nature, 59–60, 68–69, 88–90

V
vegetation, 77, 119–20, 124–27. *See also* plants
Verheyen, P., 141
vessels, 128–34, 193–94
vivisection, 58, 130
Volvox, 176
Vorticella, 110

W
whale, 162
Wiesel, J., 12
Willis, T., 42, 55, 92, 106, 129, 189; alimentary duct, 137–38; lung, 130, 136–37
Wilson, J., 24
Wren, C., 81

Library of Congress Cataloging-in-Publication Data

Fournier, Marian.
The fabric of life : microscopy in the seventeenth century / Marian Fournier.
p. cm.
Includes bibliographical references and index.
ISBN 0-8018-5138-6 (hardcover : alk. paper)
1. Microscopy—History—17th century. 2. Microscopes—History—17th century. I. Title.
QH205.2.F68 1996
502'.8'209032—dc20 96-5680